Writing Papers in the Biological Sciences

Writing Papers in the Biological Sciences

Fourth Edition

Victoria E. McMillan
Colgate University

Statistics Consultant, Robert Arnold
Colgate University

Library Consultant, Deborah Huerta
Colgate University

BEDFORD/ST. MARTIN'S Boston ◆ New York

For Bedford/St. Martin's

Developmental Editor: Jennifer Blanksteen
Associate Editor: Amy Hurd Gershman
Senior Production Supervisor: Nancy J. Myers
Marketing Manager: Karita Dos Santos
Project Management: Books By Design, Inc.
Text Design: George McLean
Cover Design: Billy Boardman
Cover Photograph: Red-Webbed Tree Frog. © Kevin Schafer/CORBIS.
Composition: Pine Tree Composition, Inc.
Printing and Binding: R R Donnelley & Sons Company

President: Joan E. Feinberg
Editorial Director: Denise B. Wydra
Editor in Chief: Karen S. Henry
Director of Marketing: Karen Melton Soeltz
Director of Editing, Design, and Production: Marcia Cohen
Manager, Publishing Services: Emily Berleth

Library of Congress Control Number: 2005931086

Manufactured in the United States of America.
1 0 9 8 7 6
f e d c b

For information, write: Bedford/St. Martin's, 75 Arlington Street, Boston, MA 02116 (617-399-4000)

ISBN: 0-312-44083-9
EAN: 978-0-312-44083-1

Acknowledgments
Screen shots for EBSCO host database search. Reprinted with permission of EBSCO Publishing, Inc. All rights reserved.

Screen shot of Tree of Life homepage. Reproduced with permission. © 2005 Tree of Life Web Project.

Preface

As a biologist who also teaches composition, I see many students who have strong backgrounds in biology but still struggle with the writing tasks and challenges unique to their discipline. Biologists need to think creatively about science, and they need to write about science clearly, accurately, and concisely. In fact, these two processes, thinking and writing, are as interdependent in science as they are in any other field. Unfortunately it is difficult to fit any extensive discussion of scientific writing into a regular semester-long biology course. I continue to hope this book can substitute for such discussion, serving as a supplemental text for courses requiring laboratory reports or longer written assignments. Although intended primarily for undergraduates who need an introduction to the aims and format of biological writing, it will also be useful for more experienced students, including master's and doctoral candidates who are preparing dissertations or manuscripts for publication. Finally, I hope this book will also find its place in the small but growing number of courses on scientific writing at the undergraduate or graduate level.

Like its three predecessors, the fourth edition is organized as a self-help manual, offering straightforward solutions to common problems and numerous examples of both faulty and effective writing. All of the book's most important features—rules highlighted with bullets, numerous examples, clear and concise explanations, and a detailed index—have been retained, including the spiral binding for easier use. The fourth edition includes expanded discussions of many topics as well as much new material.

The Introduction discusses the role of both formal and informal writing in biology and gives an overview of the diversity of writing tasks faced by professional biologists. Chapter 1, a detailed guide to locating and using biological literature, has been thoroughly revised to address the ever-changing parameters of electronic research. Coverage includes guidelines for conducting online searches, updated descriptions of a wide variety of scientific databases, and an enlarged section on evaluating Internet sources. Sections on how to read scientific papers, how to take good notes, and how to avoid plagiarism—always of concern to students—have been retained.

Chapters 2 and 3, which introduce students to handling, analyzing, and presenting data, have been enlarged and updated. Both chapters still include advice on experimental design and the use of statistics both in writing and in research.

Chapters 4 and 5 discuss the three types of writing assignments most commonly faced by biology students. Chapter 4 focuses on the *research paper,* the primary form in which professional biologists communicate original findings. Students are expected to follow this model when they prepare research projects or laboratory and field reports. In addition to an updated sample research paper, Chapter 4 includes an example of a student *lab report* as well. Chapter 5 discusses the *review paper,* which resembles the library-oriented "term paper" assignments of advanced courses. The sample paper in this chapter has also been revised.

Students often get confused about how to use literature citations and how to prepare the Literature Cited section of their papers. Chapter 6 discusses the Council of Science Editors (CSE) documentation format, including guidelines for handling electronic sources. This section has been enlarged and updated to reflect changes likely to appear in the forthcoming seventh edition of the CSE style manual. CSE format is illustrated by each of the sample papers in Chapters 4 and 5, as well as the sample research proposal in Chapter 10. New to the fourth edition is a section on APA style, the documentation system recommended by the American Psychological Association and used in psychology and other behavioral and social sciences.

Many students are uncertain about how to begin a paper or how to revise it effectively. Chapter 7 discusses the process of drafting and revising and presents the basics of each stage of the writing process—from brainstorming to composing the first draft, to final editing at the paragraph and sentence levels. To help students prepare their final drafts, Chapter 8 discusses manuscript format, provides proofreading strategies, explains how to write an Acknowledgments section, and lists symbols and abbreviations commonly used in biology.

Students often underestimate the role that writing can play in learning difficult material and in studying for examinations. Chapter 9 offers strategies for taking notes, making laboratory drawings, getting the most out of textbooks, studying for short-answer questions, and preparing for laboratory exams. Examples of poor and improved essay examination questions have been added.

Finally, the fourth edition includes updated coverage of other forms of biological writing often faced by biology students. Chapter 10 presents strategies for preparing research proposals, letters of application, résumés, curricula vitae, and both oral and poster presentations. Examples from the third edition have been updated, and both a sample PowerPoint presentation and a sample student poster have been added.

Throughout the book I have tried to address the special demands and challenges of academic assignments in the context of how and why professional biologists write. I hope this handbook will convince students that scientific writing, including their own, need not be tedious and cumbersome, but can be clear, crisp, incisive, and engaging.

Acknowledgments. I owe many people thanks for their help with the fourth edition of this book. Deborah Huerta, Science Librarian at Colgate

University and consultant for Chapter 1, provided invaluable advice and information, without which the updated material on searching the literature and evaluating Internet sources would not have been possible. Robert Arnold, of Colgate's Biology Department, served as consultant for Chapters 2 and 3 and spent many hours helping to update the tables and figures. He also helped immeasurably with the new sample PowerPoint and poster presentations, in addition to providing insightful comments on other portions of the text.

I am also deeply appreciative to Peggy Robinson, Chair of the CSE Style Manual Subcommittee, for meticulously critiquing the section on CSE documentation in Chapter 6 and for her guidance about coverage of changes in CSE style in the forthcoming seventh edition. This chapter would not have been possible without her assistance.

I continue to be grateful to former Colgate students Tara Gupta, Brian Martin, and Kristin VanderPloeg for giving me permission to adapt their writing for use in this book; to William Oostenink for use of his photomicrograph of *Smilax* in Chapter 9; and to Judy Fischer for advice about résumés and curricula vitae.

The fourth edition profited from the thoughtful comments of the following reviewers: David Caprette, Rice University; Ellen M. Dawley, Ursinus College; Terry Derting, Murray State University; Tracy Galarowicz, Central Michigan University; Michael Ghedotti, Regis University; Jean-Marie Kauth, Benedictine University; Beth Ferro Mitchell, Le Moyne College; Linda Ogren, University of California at Santa Cruz; Susan J. Rehorek, Slippery Rock University; Ellie Skokan, Wichita State University.

At Bedford/St. Martin's, I owe many thanks to Chuck Christensen, founder and former president, now retired; President Joan Feinberg; and Editorial Director Denise Wydra for their continuing commitment to this book. I also remain grateful to Steve Scipione for guiding me expertly through the first edition, and to Laura Arcari, who saw many possibilities for innovations and additions as I prepared the second edition. The third and fourth editions would not have been possible without the editorial guidance of Karen Henry, editor in chief, Boston, and Jennifer Blanksteen, development editor, and I am deeply grateful for their encouragement, patience, and many excellent suggestions. I am grateful to Mary Ellen Smith for her superb copyediting. I am indebted to Emily Berleth for skillfully overseeing the production of the fourth edition and to Herb Nolan and Carol Keller of Books By Design for their editorial and art supervision, respectively.

Over the years, many past and present colleagues in Colgate's Interdisciplinary Writing Department—Margaret Darby, Rebecca Howard, Mary Joy, Katherine Lynes, Bruce Pegg, Mary Lynn Rampolla, and others—have contributed to my growth as a writing instructor and in this way have enriched all editions of this book. Finally, I owe additional thanks to Robert Arnold for his continued encouragement and support along the way.

About the Author

Victoria E. McMillan (Ph.D., Syracuse University) teaches in the interdisciplinary writing department and the biology department at Colgate University. A behavioral ecologist who has published a number of scholarly and popular articles on animal behavior, McMillan is currently focusing her research activities on reproductive strategies in insects, dragonflies in particular.

Contents

Writing Papers in the Biological Sciences

How and Why Biologists Write: An Introduction to Biological Literature

There are several common misconceptions about the role of writing in science. One of these is that scientists don't actually do much writing; consequently, biology students, unlike students in the humanities, for example, need not worry about their writing abilities. Another misconception is that scientific writing is, at best, dry and cumbersome; at worst, it is tedious, pedantic, and impenetrable. Scientists are rarely credited with the ability or the motivation to write with creativity and flair, and the diversity of writing tasks faced by most biologists is often underestimated.

This book, I hope, will help dispel such misconceptions. Science is a collective enterprise: its growth depends on the work and insight of many individuals and on the free exchange of data and ideas. Biologists *must* be effective writers because no experiment, however brilliant, can contribute to the existing fund of scientific knowledge unless it has been described to others working in the same field. For this reason, the primary aim of scientific writing is to communicate—as clearly, accurately, and succinctly as possible. This goal is not easy to achieve, and examples of stilted, jargon-ridden papers do appear in the scientific journals. However, there are also many examples of good, even brilliant, scientific writing whose authors have managed to translate complex data and theory into clear, straightforward, illuminating prose. Such works have their own kind of grace and elegance.

Most biologists routinely engage in a variety of writing tasks. Their professional work is typically published in scientific journals and is aimed at a specialized group of readers: generally other scientists with some background in the particular subject area. Biologists write two major types of papers for such journals: the research paper and the review paper.

The *research paper* reports original findings from a field or laboratory investigation that may have extended over many weeks, months, or even years. It is generally divided into six distinct sections: Abstract or Summary, Introduction, Materials and Methods, Results, Discussion, and Literature Cited (sometimes called References). Although editorial requirements vary from journal to journal and authors must conform to those specifications, most scientific papers follow the same basic organization. The format has been dictated by a long history of printing and publishing traditions that reflect the logic and elegance of the scientific method.

The format of a research paper reflects a scientist's obligation to make his or her assumptions clear, the methods repeatable, and the interpretations clearly separate from the data or results. Although it may contain lengthy descriptive data or complex statistics, a research paper is at heart a well-structured *argument* founded on supporting evidence. It serves as a vehicle for presenting one's own findings and conclusions and for arguing for or against competing hypotheses. If independent investigators use similar methods but come up with different results, then the conclusions of the original study may need reevaluation or modification. In any case, our understanding of the subject will eventually be expanded.

In more practical terms, research papers serve as a principal means by which a particular scientist's contributions are evaluated by his or her peers. As professional biologists know all too well, it is not easy to turn raw data into a published research paper in a journal of one's choice. Most papers have gone through multiple drafts and often extensive revisions before appearing in their published form. Typically, the author solicits comments from his or her colleagues before submitting the paper for publication. In addition, the journal editor sends the manuscript to several reviewers — other scientists in the author's field — who read it critically and recommend that it be either accepted or rejected. Even if accepted for publication, a manuscript usually needs some changes, often major ones, before finally finding its way into print. The entire editorial process takes many months to complete.

Several other, related forms of writing can also convey a biologist's data or ideas. In addition to full-length research papers, many journals also publish shorter reports of original findings. They often appear in a separate section titled Research Notes, Communications, or the like. The organization of such works is similar to that of research papers, although certain sections may be streamlined or abbreviated to conform to shorter manuscript specifications.

A *conference presentation* is a paper presented either orally or in poster format at a formal meeting of scientists interested in a particular field of study. Conference papers present unpublished research, often work in progress; biologists depend on them to learn about the current activities of their colleagues. The abstracts of presentations are usually distributed in printed form to all people registered for the conference; the proceedings of major symposia are listed in major indexes and abstracts and can be obtained through research libraries. Typically, authors of conference talks and posters publish their work as journal papers as soon as possible.

A *research proposal,* or grant proposal, is a carefully organized plan for future research. Biologists submit such proposals to particular committees or agencies — for example, the National Science Foundation—in the hope of obtaining funding for their work. Biology students may also write research proposals in anticipation of independent coursework or Ph.D. research. Because they suggest, rather than present, original studies, research proposals may have lengthy introductions that review existing literature on the topic and provide a rationale for the new work being proposed. Proposals also include the methodology and projected costs for equipment, research assistants, travel, or other expenses. A results section is lacking, of course, although the introduction may showcase the author's previous findings, particularly if they were obtained with funds from the same agency.

In contrast to research papers, conference presentations, and research proposals, a *review paper* is a journal article that synthesizes work by many independent researchers on a particular subject or scientific problem. By bringing together the most pertinent findings of a large number of studies, a review paper serves as a valuable summary of research. Although it does not present the writer's new discoveries, it does reflect his or her painstaking review of the literature in a defined field. Moreover, a good review not only summarizes information but also provides interpretive analysis and sometimes a historical perspective. Reviews vary in aims, scope, length, and format, but they all include a relatively lengthy reference section. Journal editors sometimes invite prominent experts to write reviews of their particular fields, since the ability to give an authoritative overview of a subject usually develops with experience. Whether solicited or unsolicited, review papers still must conform to journal specifications, and their authors receive feedback from editors and reviewers before final publication.

In addition to writing for specialized scientific audiences, many biologists also write extensively for readers with little background in science. For example, magazines such as *Natural History,* published by the American Museum of Natural History, and *Smithsonian,* published by the Smithsonian Institution, feature a wide variety of essays and articles written by experts for nonspecialists. Many scientists, in fact, take great pleasure in transforming their professional work into writings that reach out to the general reader. In addition to magazine submissions, some biologists write books on natural history, health, the environment, or the practice of science; others may compose poetry or write novels. Still others combine science with their involvement in teaching through the writing of textbooks (like this one) or articles in journals such as *Journal of College Science Teaching* or *The American Biology Teacher.*

Biologists who teach at colleges and universities routinely produce a variety of other written products: lectures and course syllabi, laboratory and paper assignments, examinations, letters of recommendation for students and colleagues, memos and departmental reports, and, at times, their own résumés and letters of application for new academic positions. Finally, remember that most *formal* writing is preceded and molded by *informal* writing in the form of notes, outlines, journal entries, laboratory or field data, and communication with colleagues through letters, professional newsletters, and e-mail and Internet discussion groups. In short, good writing skills are as indispensable to biologists as they are to scholars in other academic disciplines.

CHAPTER 1

Locating and Using Biological Literature

Review paper assignments, research projects and proposals, and many laboratory reports will require you to immerse yourself in biological literature. All scientific endeavors, including your own, acquire their fullest meaning and significance when viewed in the broader context of what other scientists, past and present, have thought and done. Therefore, to be a successful biology student and, with time, an effective biologist, you will need to develop competent information-literacy skills. In other words, you will need to know how to locate the scholarly work of others, how to critically evaluate these sources, and how to incorporate this information effectively into your own writing. The guidelines below will help you get started.

SEARCHING THE LITERATURE

■ Understand the difference between primary and secondary sources.

In the sciences, *primary sources* are reports of original findings and ideas. These generally take the form of peer-reviewed research papers in scholarly journals and are directed toward a scientific audience. Thousands of scientific journals are published around the world. Some, such as *Science* and *Nature,* report findings from diverse branches of science; others, such as *Neurology, Phycologia,* and *Journal of Bacteriology,* are focused more narrowly on particular subdisciplines. Research papers in journals usually deal with relatively specialized topics—for example, "Osmoregulation in the Freshwater Sponge, *Spongilla lacustris*"; "Pelagic Carbon Metabolism in a Eutrophic Lake during a Clear-water Phase"; or "Facilitated Nuclear Transport of Calmodulin in Tissue Culture." Sometimes a research paper reappears later in book form, as part of an edited collection of noteworthy articles on a particular subject. Although many journal articles are written in English and may be accompanied by an abstract in another language, such as French, German, or Japanese, others may be written in a language other than English and accompanied by an Abstract in English.

Aside from research papers, other primary sources are conference papers, dissertations, and technical reports of government agencies or private organizations. Your clue to recognizing primary literature is usually the presence of a Materials and Methods section, or at least some mention of methodology, as well as a Results section, typically with accompanying tables or figures. These components indicate that the author is presenting new data and ideas.

Secondary sources are more general works that are *based* on primary sources. Many are written for readers with little knowledge of the sciences. These include books on natural history, the environment, and other scientific subjects, as well as articles in magazines such as *Discover, Natural History, Smithsonian, American Scientist, Audubon, Scientific American,* and *National Wildlife.* Although many secondary sources are written for both scientists and nonscientists, others attract a more specialized audience; examples are scholarly books (monographs) and review papers that summarize and interpret the primary literature in a particular subject area. Finally, note that secondary sources, like primary sources, may appear in both print and digital formats.

■ Start by consulting general references.

Before plunging into the specialized primary literature on your topic, make sure you have a solid background and some sense of your aims and scope. For example, if you are writing a paper on regeneration in planarians for an invertebrate zoology course, you might begin by reading all relevant sections of your own course text. Next, you might search for other general books on the biology of flatworms or on regeneration in invertebrates. Such references will give you some grounding in the topic and may contain extensive specialized bibliographies. Scientific encyclopedias also make good introductory reading and usually contain additional references.

Your library's card catalog will help you locate these and other general sources on your subject. Although diverse interfaces abound, all online library catalogs provide the same basic functions: author, title, keyword, subject, and combination searching as well as links to other library resources and services. Keyword searches allow broad inquiries concerning accessible materials about a given subject. Subject searching, based on controlled vocabulary, is more precise. Usually a combination of keyword and subject vocabularies constitutes a successful search strategy.

In addition to general background sources, popular science magazines provide accessible introductions to difficult biological material. Their articles give up-to-date overviews of particular subjects and may mention names of important researchers. Their nonspecialist language and approach will prepare you for the more specialized vocabulary of scientific journals—which ultimately should be your major reading material. Many authors of nontechnical articles have also published scholarly papers on the same topic, and later you may wish to locate those papers, too.

Review papers provide excellent beginning reading. Although written for scientific audiences, reviews may not require as much background in the specific subject area as do the more specialized research papers they discuss. Review papers give you some historical perspective, summarize the contributions of influential researchers, and provide a sense of the important scientific issues being addressed. Their Literature Cited section will be a rich source of primary literature to consult once your own focus has narrowed. Published reviews are also appropriate models to consult if you are preparing your own review or term paper (see Chapter 5).

Although some journals, such as *Genetics,* specialize only in research papers, many other publications are devoted exclusively to reviews. The latter include *Annual Review of Ecology and Systematics, Advances in Ecological Research, Recent Progress in Hormone Research, Annual Review of Genetics,* and numerous others. Still others, such as *Biological Bulletin,* publish mostly research papers but also feature occasional review papers.

Finally, do not overlook the usefulness of scholarly monographs, which provide a more extensive overview of the literature than is possible in a review paper and serve as excellent introductions to more specialized scientific topics. Your reference librarian can assist you in locating such resources.

■ Learn to use scientific databases.

The Internet and World Wide Web are rapidly changing both scientific literature and access to it. Most libraries provide database access to primary and secondary literature in journals, as well as to monographs, dissertations, conference proceedings, documents, and other reports. Each database has its own method of organization; you will need to learn the different protocols for electronic databases by accessing their detailed Help screens. In either case, the most efficient strategy is to consult your reference librarian.

Following is a selection of indexes and abstracts most useful to biologists:

Either *General Science Abstracts* or *Academic Search Premier* is an excellent database in which to begin your literature search. Both provide access to a wide variety of leading science journals and magazines, including those covering biology, environmental issues, and health. Many full-text articles are accessible in PDF or HTML formats.

Biological and Agricultural Abstracts, slightly more specialized, provides access to information in environmental and conservation sciences, agriculture, forestry, veterinary medicine, and applied and field biology. It lists articles in selected general journals (for example, *Heredity, The American Naturalist,* and *Ecological Monographs*) as well as in more specialized periodicals (*Crop Science, Tropical Agriculture, Ecological Entomology,* and *Animal Science*). Beginners find it especially useful because it includes many periodicals devoted solely to reviews.

AGRICOLA, the National Agricultural Library's database, covers a greater range of similar subjects and also provides access to more specialized books, journal articles, and documents. Subjects include agriculture, natural resources, aquaculture, plant and animal sciences, forestry, entomology, food and nutrition, environmental issues, and veterinary medicine.

CAB International is a leading international database for agricultural topics, including biotechnology, developing nations' agricultural concerns, and agricultural genetics, as well as related fields of inquiry.

Biological Abstracts (BIOSIS PREVIEWS or Basic BIOSIS) is the largest life sciences journal literature database. It includes a companion database, *Biological Abstracts/RRM,* which indexes reports, reviews, and meetings. In addition, the same vendor offers the venerable database *Zoological Record,* originally published by the Zoological Society of London and the British Museum. These three databases are available in a variety of formats with an interface with *Web of Science (Science Citation Index).* This arrangement will allow links to full-text documents available through databases, provided your library subscribes to these services. All these databases provide full information on references, including abstracts.

Cambridge Scientific Abstracts (CSA) also offers a number of fine biology databases. In its *Biological Sciences* database are numerous subfiles for such fields as biology, genetics, ecology, pharmacology, medicine, and bioengineering. CSA also maintains *Environmental Sciences and Pollution Abstracts, Oceanic Abstracts,* and databases covering genetics, microbiology, aquatic sciences, agriculture, and many specialized subfields. These databases connect to selected Web sites as well as to a list of references with abstracts. They, like other online databases, may feature connections with your library's online catalog and/or interlibrary loan programs.

If animal or human behavior interests you or if you are working on psychiatric or psychological topics, *PsycINFO,* the immense database from the American Psychological Association (APA), provides citations for journal articles, book chapters, books, and dissertations. International in coverage, *PsycINFO* indexes materials from the nineteenth century to the present. A companion database, *PsycArticles,* has full-text connections to APA publications.

MEDLINE, the National Library of Medicine's bibliographic database, is an indispensable tool for locating medical, veterinary, dental, and other health-related literature. *PubMed* includes MEDLINE and OLDMEDLINE (journal articles from 1950-1965) as well as other research links and services. Provided that your library has subscribed, you may be able to access some full-text articles. *PubMed* also offers a "related articles" feature allowing you to access an additional set of references related to a particular document. This feature is also offered by other databases such as *Scifinder Scholar* or *Web of Science.*

Another comprehensive medical database is EMBASE, which indexes international literature on medicine and related subjects such as paramedical professions, drugs, optometry, public health, gene technology, hospital management, and other medical specialties.

The *Cumulative Index to Nursing and Allied Health Literature* (CINAHL) is a major access tool for sources on nursing and allied health professions. Subjects range from respiratory therapy and cardiopulmonary technology to health education, medical records, and many others.

In addition, the Gale Group *Health & Wellness* database provides wide popular coverage of a variety of health and medical topics. This database provides mainly full-text access to reference materials, books, journals, pamphlets, and publications from medical and government organizations.

Scifinder Scholar and STN provide online access to *Chemical Abstracts,* the world's largest database for accessing chemical and biochemical literature. This database is extremely helpful for locating research in the areas of biochemistry, genetics, environmental chemistry, pollution, and biomedicine. *Medline* is embedded in *Scifinder Scholar,* so you may search both databases simultaneously. If your library subscribes to the American Chemical Society full-text journals series, you will be able to use *Scifinder Scholar* to access articles.

GEOREF is the key to the literature of earth sciences and geology, including paleontology; *Geobase* indexes geology, geography, and ecology. Both databases are available in a variety of formats and are always worth searching for topics in environmental sciences.

GPO Monthly Catalog is the U.S. Government Printing Office database, an extraordinary resource accessing the vast array of items generated by federal agencies. Biologists may be particularly interested in documents about agriculture, energy, the environment, and health, as well as hearings concerning relevant biological issues. Of special note is Science.gov (http://www.science.gov), a portal for science databases, bibliographies, and links to resources provided by U.S. government departments and agencies. In addition, various agency Web sites have internal search engines for linking to their online resources, such as those for the NIH, the EPA, Fish and Wildlife Service, NOAA, or the CDC.

Web of Science, an extensive multidisciplinary science database, provides coverage far more extensive than other broad coverage services. This unique database allows you to identify articles that cite a particular author and a particular document. Included in each record is the article's Literature Cited section. Like *PubMed* and *Scifinder Scholar, Web of Science* provides a related articles feature.

Web of Science is a search tool with multiple uses, available through libraries that subscribe to this service. If you already know of one paper relevant to your topic, you can use the reference to that paper to locate other, more recent papers by authors who have mentioned (or cited) it. Chances are that those authors have written about the same subject. First, look up the known paper (say it's by J. L. Jones in 1990). You may find one or more references to authors who have cited Jones (1990) in their own work. Full bibliographic information about these papers will be provided, and you may also be given the abstracts. You can then go to the appropriate journals or online access to retrieve the papers themselves.

You can use *Web of Science* in several other ways. For example, if you suspect that J. L. Jones has published other papers since 1990, you can find them listed under Jones's name by doing a simple author search in the database. As another example, if you know that scientists at Northwestern Medical School in Chicago are particularly interested in the contraceptive effects of breast-feeding, you can locate articles on this subject originating at this institution. Finally, if you have thought of one or more keywords pertaining to your topic, you can use these to locate other relevant sources.

Web of Science embeds certain "current awareness sources," also known as *Current Contents,* covering life sciences, social and behavioral sciences, medicine, agriculture, biology, and environmental sciences. Many other current awareness publications include *Current Opinion in Cell Biology, Current Opinion in Genetics and Development, Trends in Biochemical Sciences, Evolution,* and *Trends in Neurosciences.* Your library may subscribe to full-text journal services to these publications.

BioOne and *JSTOR* provide full-text access to biology and other science journal collections. *BioOne* offers an aggregate of bioresearch titles published by small professional societies and noncommercial publishers. Supported by a collaboration of scholarly interests, including the American Institute of Biological Sciences, *BioOne* is accessible as a stand-alone database using CSA's Illumina interface.

JSTOR archives full-text scholarly collections in core fields such as biology, botany, ecology, and general science. *JSTOR* puts an embargo on new titles, so the most recent issues of journals are not available. Various search levels are utilized for specific recall of known articles. You may also conduct general subject, author, or title searches.

"Gray literature" of interest to biologists consists of unpublished research in the form of technical reports, working papers, and some conference proceedings. Prior to the Internet, gray literature was considered elusive; now it can be found on institutional or government Web sites. Online directories and guides are available to help you access gray literature.

Many of the databases outlined in this section have clickable connections to bibliographic utilities such as Refworks, EndNote, Procite, or Reference Manager. These programs allow you to store your references online in folders and to print out your Literature Cited section in a particular documentation format (CSE, APA, and so on). Even if the connection from the database to the utility is not available, there are work-arounds explained in the Help screens to assist you.

Do not expect any one database to provide all the references needed to achieve your research goals. Paper indexes are still highly valuable for historical research, since they include older sources that may not appear in online databases. Consult your reference librarian for instruction about the various features of each database so that you can use these resources effectively and efficiently.

■ Plan your search vocabulary carefully.

As you begin your literature review, you'll need to build a vocabulary that, although precise, still allows the broadest flexibility in your search. Keywords and authors' names are the chief means of accessing information in databases. Use your online library catalog to identify precise subject terms that direct you to the most appropriate materials on your topic. If you are having little success in researching a particular topic, ask a reference librarian for help in expanding or revising your search terms. Be sure to consult the online thesaurus for the databases you are using. Note that acronyms such as AIDS or abbreviations such as DNA are often useful search terms. Be prepared for alternative spellings: for example, the British *colour* instead of the American *color,* or *behaviour* rather than *behavior.*

Note that truncation features or wildcards are available for most databases; these allow you to search for the singular and plural forms of a term simultaneously or to access terms containing particular roots. As an example, suppose you are interested in estuaries and decide to search BIOSIS Previews. It may seem logical to start with *estuary* as a search term. However, by using the appropriate truncation symbol (for example, *) to make *estuar**, you may retrieve a larger number of potentially useful references. The root *estuar** will locate documents pertaining to estuary, estuaries, and estuarine. You can further refine your search by using additional keywords in combination with *estuar**.

SAMPLE DATABASE SEARCH

Science literature databases provide you with the means to search efficiently for scholarly materials on a topic. For example, suppose you are interested in bird songs. You might begin your search with a general comprehensive database such as Academic Search Premier. Beforehand, of course, you should have done some preliminary research on the subject to provide yourself with background information and a list of synonyms with which to conduct a search.

As a first step, open Academic Search Premier and type the keyword "bird song" into the search field. Note that Academic Search Premier, like most databases, allows you to select the field within which you want to search. The database suggests "Default Field," which means the software will search for your keyword, "bird song," in the author, subject, keyword, title, and abstract fields. This unrestricted search is very broad and should retrieve all articles in the database containing your keyword.

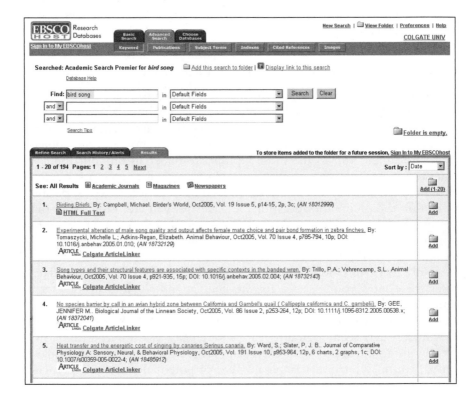

Your search has retrieved 194 items. Not only are there too many items to explore efficiently, but also some may be inappropriate, such as #4. Therefore, to be effective you will need to restrict your search. There are several ways to narrow this search. The easiest and perhaps most effective tactic would be to limit the keyword "bird song" to the title field.

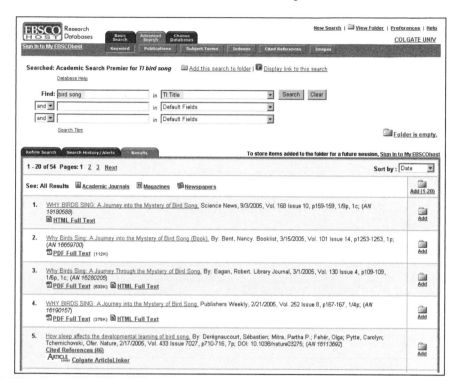

Now you have a subset of your original search that has retrieved 54 items with "bird song" in the title of an article. Although not every article will satisfy your criteria, you have retrieved one-third of the original search and have acquired a set of helpful citations.

You may wish to make this search even more effective by adding another term. For example, if, after thinking more about your topic, you decide that you are most interested in how birds learn their songs, you might redo your search by adding the term "learning" and leaving that term in the default fields.

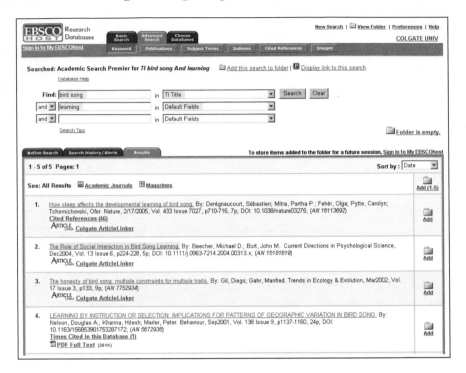

You have now retrieved five articles that are very specific. Note that in this database and many others you may retrieve a full-text PDF file, as well as metadata about these items.

After reviewing your results, you might wish to broaden your search to retrieve additional materials. This process is easily accomplished. Open the screen for the metadata about article #2, "The role of social interaction in bird song learning," by clicking on the article title.

Here is the metadata screen describing the article. You now have the abstract as well as the title, authors, source, and subject (indexing) terms. Thus, you have enough information to decide whether or not to procure the article itself. The article may be available online, in a printed journal in your library, or through your library's interlibrary loan service.

Note the different subject terms by which this article was indexed (including the term "birdsongs" as a single word). You might have discovered these terms earlier by consulting the online thesaurus of subject terms used by this database for indexing purposes. (See the clickable button at the top of the screen.) Also note that the authors' names may be of further use to you, since the authors' expertise may be in your subject area. Their names are searchable in this database and other, more specific, subject databases.

If this paper interests you, you might click on an author's name for additional articles about this subject.

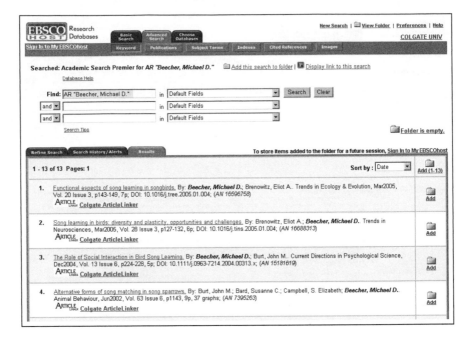

The database has now placed the principal author's name in the search window and is displaying his articles indexed in Academic Search Premier. As you can see, you have discovered a researcher whose published materials reflect an interest in bird songs and learning. However, this author's research may or may not be exactly what you want to study. Recall that the database uses the keyword "birdsongs" rather than your initial keyword, "bird song." That slight modification may or may not be significant in your search; however, it may be worthwhile to investigate using the keyword "birdsongs." To do this, you can go back to the previous screen (p. 15) and click on "BIRDSONGS" in the list of subject terms, or you can type DE "BIRDSONGS" into the search field, as shown in the next screen on page 17. ("DE" denotes "descriptor" in your search.)

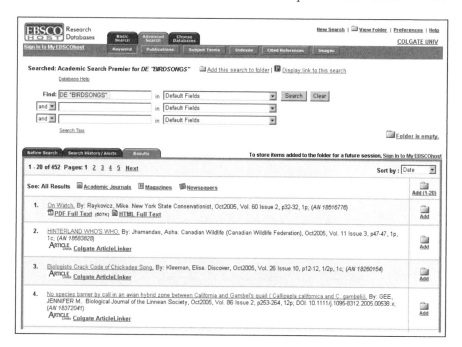

This new search is actually larger than your original search and may provide you with a better set of results as long as you narrow the scope. You may accomplish this by inserting the term "learning" in the second search window, which will pair "learning" with "birdsongs" as a descriptor.

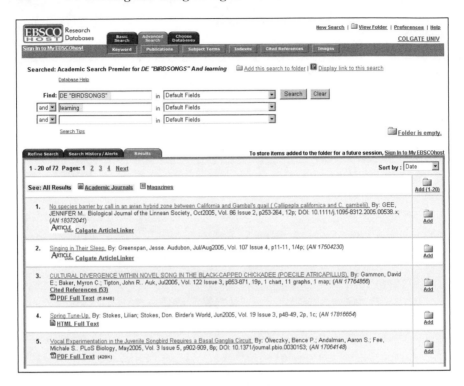

Your next results are a set of items reflecting this new search strategy. Depending upon your project, you may want to further narrow the set by clicking on "Academic Journals," which results in articles found only in scholarly, peer-reviewed sources.

At this point, you may wish to retrieve full-text articles and collect other citations that appear promising. Most databases will allow you to mark desired items and print only those items. In this database you click on the folder icon next to each item to mark your selections, and then click to open the folder containing your choices.

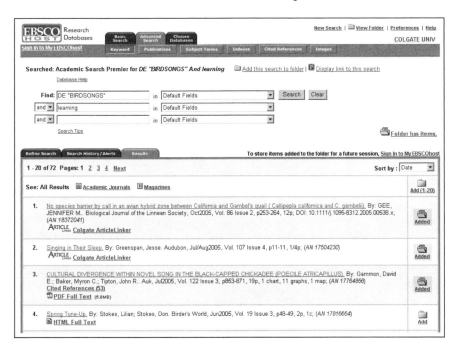

When you click on the folder containing your marked selections, you will be offered several options for saving your choices. Most databases allow you to save your results to a disk or to a bibliographic utility. You may also print your list of references or e-mail it to yourself or to your instructor for confirmation that you are locating appropriate materials, as shown in the next screen on page 20. There may be other features, as well, that allow you to store your search logic for retrieval or updates.

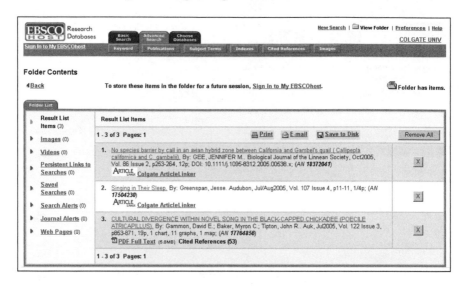

Not counting the time it takes you to review the various items you have retrieved, it should take only 10 to 15 minutes for an effective search. Again, some background knowledge of your topic and a little advance planning are critical for success.

■ Use the Literature Cited or References section of relevant papers to find additional primary sources.

Once you are comfortable gathering primary references, one of the most efficient ways to expand your list of references is to take advantage of the experience of established researchers. Each time you find a useful paper, study its Literature Cited section carefully. In it you may discover some additional titles that are relevant to your work. The author's comments about these sources in the Introduction or Discussion section of the paper may give you further clues about their contents, perspective, or significance. Also, you can enter these sources in the "cited references" search page in the *Web of Science* to locate still more papers on your topic.

USING THE INTERNET

■ Investigate information resources on the Internet.

Using the Internet, you will find a huge variety of resources, including specialized data files; electronic journals and newsletters; preprint, text, and software archives; and special-interest discussion groups and mailing lists pertaining to seemingly every conceivable subject—from potato growing to camels, from paleoclimatology to bird banding. Not all publicly available databases are bibliographic; there are also databases for gene sequencing (GenBank), protein data (Protein Data Bank), and other specialized but accessible sources for statistics.

Internet resources constitute a bewildering array of connections to academic, commercial, governmental, and private users and services; moreover, these resources are constantly changing. To use the Internet or, in particular, Web sources, you must devise a focused search strategy. Therefore, you need enough background knowledge to make a list of important concepts related to your topic. For example, suppose you are interested in primate behavior and have decided that the following general and specific keywords are most significant for your search:

primates ⇒ lemurs ⇒ sifakas ⇒ *Propithecus verreauxi*

behavior ⇒ families ⇒ play

How can you best combine these terms for the most efficient and effective search strategy? First, you need to know something about search engines. A search engine is a program that runs on a graphical browser and allows you to use key terms to conduct a search on the Web. You are probably already familiar with Google. Some search engines—such as *Yahoo!*—search their own carefully selected database. Other search engines, such as *AltaVista,* look for specific terms, noting the frequency with which a term occurs on a particular Web site. Another approach, the metasearch engine, such as *Dogpile*, searches the databases of other search engines.

Search engines employ tools for searching that are already familiar if you have used an online library catalog. For example, in the Advanced Searching option, many browsers feature the Boolean operators AND, OR, NOT, and NEAR. If you were searching the Web for lemurs and behavior as suggested above, you could limit or increase your results by using a combination of keywords and Boolean operators. You might expand your search using the combination (lemurs OR sifakas) AND play. By "nesting" the keywords connected by OR within parentheses, you are instructing the program to first look for the terms "lemurs" or "sifakas" and then look for documents and Web sites that also include the word "play" with either "lemurs" or "sifakas". Because your nested terms are connected with the operator OR, your search will include Web sources with either "lemurs" and "play" or "sifakas" and "play." If you wish to limit your search further, you can choose specific keywords, such as the species of lemur ("*Propithecus verreauxi*") placed between quotation marks to signal a search for this exact phrase and combine this phrase with "play behavior." In addition, you may use the operator NEAR in many search engines to indicate that two words may be separated but in close proximity. For example, the search phrase "lemurs NEAR play" might allow you to pick up a Web site entitled "Play Behavior in Lemurs." Truncation symbols are a way to broaden your scientific search just as you might in online catalogs and literature databases.

Search engines provide valuable tips on specific usage if you click on the Help button. For an excellent nitty-gritty description of search engines and searching techniques, see BARE BONES 101: A Basic Tutorial on Searching the Web (http://www.sc.edu/beaufort/library/pages/bones/bones.shtml).

■ Evaluate Web sources carefully.

Even the most refined search strategy will yield an overwhelming number of Web sites. Of course, you must select the most relevant references to use for your paper. Generally you will need to focus on primary rather than secondary sources. Many print, peer-reviewed journals have digital counterparts (for example, *Molecular Genetics and Metabolism*). Some primary journals, such as *Evolutionary Ecology Research,* have no print equivalent and appear *only* online. Original research also appears on many organizations' Web sites, such as that of the Centers for Disease Control and Prevention (CDC) at <http://www.cdc.gov>. The CDC publishes morbidity and mortality statistics, epidemiology reports, and many other kinds of invaluable data.

Also of interest to biologists are the open-access/open-archives initiatives. For example, in the open-archives PubMed Central (http://www.pubmedcentral.nih.gov/), research generated with National Institutes of Health (NIH) financial support is freely available to the community. Public Library of Science (http://www.plos.org) and its subsidiaries, PLOS Biology, PLOS Medicine, and others provide full-text research articles that also have freely available content.

Many Web sites are clearly secondary sources and may provide substantial background information, but you will still need to evaluate their relevance in much the same way you would evaluate any other document. Begin by closely examining the URL (uniform resource locator), as in the following example:

<http://www.bioone.org/>

    1   2   3

1. Protocol
2. Authority or Address
3. Domain

Here, *bioone* is the name of a collection of important biosciences journals. The site resides in a domain labeled *organization* (".org"), indicating that it is not registered as an educational site (".edu") or as a commercial site (".com"). Further examination of the home page should reveal the purpose of the organization. Common domains are listed below; note that each identifies the basic intent of the item.

COMMON DOMAIN EXTENSIONS

Domain	Category	Example
.ac	academic	<http://www.cam.ac.uk/> University of Cambridge The .uk indicates country of origin (United Kingdom)
.com	commercial	<http://www.discovery.com/> The Discovery Channel
.edu	education	<http://www.evergreen.edu/> The Evergreen State College
.gov	government	<http://www.cdc.gov/> Centers for Disease Control and Prevention
.mil	military	<http://www.defenselink.mil/> U.S. Department of Defense
.net	Internet	<http://www.nyct.net/> New York Connect (Internet provider)
.org	organization	<http://www.amnh.org/> American Museum of Natural History

The domain tells you about the Web site's sponsor. We assume that sites with education, government, or organization domains are not attempting to sell something and are not presenting biased information. Of course, this is not entirely true.

Web sites, therefore, span a broad range—from bibliographies and scholarly articles to conference programs, book reviews, newsletters, and materials for children. However, not all of these findings will be relevant for your paper, nor will they necessarily stay retrievable, since Web sites are not always stable. How do you evaluate a mixture of results, given the mercurial nature of Web offerings? Your best strategy is to consider the following criteria:

- *Authority.* Who is the author? Does that person do research at an educational institution or a sponsored government agency? What is the author's purpose? Can you find his or her name in any of the biology databases? Have other biologists cited this author's work?

- *Sponsorship.* Who is the sponsor of the Web site? Is it a commercial or not-for-profit enterprise? Does that matter for your search? Is the argument or message of the Web site balanced in tone, or does it appear biased? Are the sources reputable? Are there advertisements?

- *Content.* Did you find reliable scientific information? How do you know? Are the author's conclusions supported by references or backed by an organization? Is there a print equivalent of this work?

- *Currency.* When was the Web site established? How recently was the Web site updated? How new is the information? Is the coverage complete?

- *Accuracy.* Are there typos or misspellings? Jargon?

As an excellent example of a useful and authoritative Web site, consider the Tree of Life project (http://tolweb.org/tree/phylogeny.html), which provides a wealth of information on evolution and biological diversity (see p. 25). The purpose of this site is immediately available by clicking the "about" link at the top of the page or the one below Darwin's picture. There is a search engine embedded in the Web site for quick browsing. The graphics provide quick entrée to major kingdoms. The authority, sponsorship, and currency are obvious from the opening screen. The site is clearly stable as seen from the dates of presentation and frequent updates. The content is described as international contributions from biologists; therefore, we assume the content is accurate.

The following Web sites may also be helpful to you in evaluating Web sources:

"Evaluating Web Sites: Criteria and Tools"
<http://www.library.cornell.edu/olinuris/ref/research/webeval.html>

Evaluation criteria from "The Good, The Bad & The Ugly: or, Why
It's a Good Idea to Evaluate Web Sources"
<http://lib.nmsu.edu/instruction/evalcrit.html>

"How to Evaluate an Internet-Based Information Source"
<http://biome.ac.uk/guidelines/eval/howto.html>

"Evaluating Information on the Internet"
<http://www.lib.purdue.edu/itd/techman/evaluate.html>

As the Tree of Life Web site illustrates, the Web makes a wealth of science images and other graphical resources accessible to students. Most search engines now provide keyword-accessible image browsing, but there are also Web sites specializing in science imagery. Here are samples:

Big Picture Book of Viruses
<http://www.tulane.edu/~dmsander/Big_Virology/BVHomePage
.html>

Botanical Society of America Online Image Collection
<http://images.botany.org/>

CELLS Alive!
<http://www.cellsalive.com/>

The Visible Human Project®
<http://www.nlm.nih.gov/research/visible/visible_human.html>

The National Science Digital Library (NSDL, http://nsdl.org/) also provides an index to science images on the Web. The library is a portal to

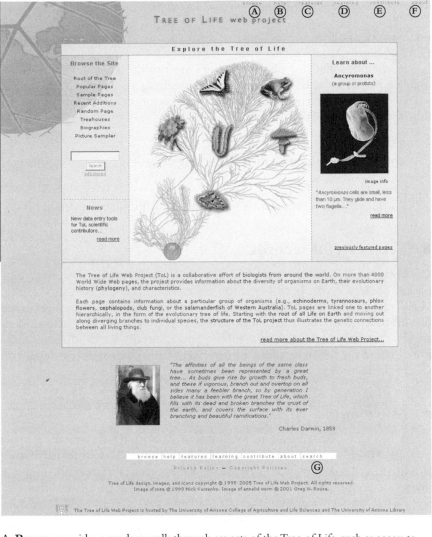

TREE OF LIFE web project Ⓐ Ⓑ © Ⓓ Ⓔ   Ⓕ

Explore the Tree of Life

Browse the Site

Root of the Tree
Popular Pages
Sample Pages
Recent Additions
Random Page
Treehouses
Biographies
Picture Sampler

Search
advanced

News

New data entry tools
for ToL scientific
contributors...
read more

Learn about ...

Ancyromonas
(a group of protists)

image info

"*Ancyromonas* cells are small, less
than 10 µm. They glide and have
two flagella...."

read more

previously featured pages

The Tree of Life Web Project (ToL) is a collaborative effort of biologists from around the world. On more than 4000
World Wide Web pages, the project provides information about the diversity of organisms on Earth, their evolutionary
history (phylogeny), and characteristics.

Each page contains information about a particular group of organisms (e.g., echinoderms, tyrannosaurs, phlox
flowers, cephalopods, club fungi, or the salamanderfish of Western Australia). ToL pages are linked one to another
hierarchically, in the form of the evolutionary tree of life. Starting with the root of all Life on Earth and moving out
along diverging branches to individual species, the structure of the ToL project thus illustrates the genetic connections
between all living things.

read more about the Tree of Life Web Project...

"The affinities of all the beings of the same class
have sometimes been represented by a great
tree... As buds give rise by growth to fresh buds,
and these if vigorous, branch out and overtop on all
sides many a feebler branch, so by generation I
believe it has been with the great Tree of Life, which
fills with its dead and broken branches the crust of
the earth, and covers the surface with its ever
branching and beautiful ramifications."

Charles Darwin, 1859

browse help features learning contribute about search

A. **Browse** provides a random walk through aspects of the Tree of Life, such as access to the tree, model pages, interactive instruction, biographies of scientists, and a search engine for the site.

B. **Help** explains the organization and navigation of the Web site and provides a sitemap, frequently asked questions section (FAQ), and contact information.

C. **Features** denotes special functions that allow you to customize the site, participate in discussions, or download the entire Tree of Life structure. Links are provided to various topics such as Biodiversity, Images, Evolution, Education, and Databases.

D. **Learning** encourages interested participants to contribute to the Tree of Life by creating "treehouses" with movies, games, text, audio, and other media. There are guides for entering material and a toolkit for assistance.

E. **Contribute** contains the treehouses and also core contributions of scientists, articles, notes, and images.

F. **About** covers the history of the project, policies, quality assurances, staff, and news. Even a model citation for the Web site is provided.

G. **Search** may be a simple search or a more advanced search that employs a Google engine to search by taxa, text, or images. The pages that display various groups have a variety of features: introduction, characteristics, systematics, species lists, life history, behavior, distribution, extensive references, and images. Each entry is dated, signed, and copyrighted.

25

other science collections such as Best Links about Science, Mycology Net, Flora of North America, and many others.

Remember that just because images are readily available on the Web, they are not necessarily available for legal use. In fact, copyright use of such works is strictly governed. For further clarification of copyright issues, the following Web sites offer guidance.

A Visit to Copyright Bay
<http://www.stfrancis.edu/cid/copyrightbay/>

Crash Course in Copyright
<http://www.utsystem.edu/OGC/IntellectualProperty/cprtindx.htm>

■ Keep careful records.

Your research will be more efficient and you'll avoid unnecessary repetition of steps if you keep a record of the search process as you go along. Write down, for example, which databases you consulted, what keywords or authors' names you used, how far back in time your search extended, and so on. Also, record *complete* bibliographic details about all potentially useful sources, including those on the Web. See Chapter 6 for details on what information you should include.

■ Allow ample time.

It is easy to underestimate the amount of time it takes to do a thorough literature search. Choosing a preliminary topic takes time; narrowing and focusing it takes even longer. Successful library research is not simply a matter of accumulating many references as quickly as possible—this will just make you feel overwhelmed. Stop at intervals to read or at least skim your best sources, and use them to help you refine your purpose and scope so that your search becomes increasingly focused and more specific.

Remember, too, that your literature search will probably turn up many promising titles that are not found in your own library. Here you have options. One is to use the interlibrary loan system: at your request, your library can usually borrow a book you need or obtain a copy of an article from a journal carried by another institution. However, this process may take several days or even weeks, especially in the case of more obscure materials, so you need to plan ahead. Be selective in your requests, since an interlibrary loan service costs time and money. Another option is to write or e-mail authors directly and request reprints of their papers. As with interlibrary loans, allow plenty of time.

For modest assistance with research timelines, you might turn to a device such as the Assignment Calculator created originally by the University of Minnesota Libraries. See the following Web site for such a calculator, adapted by the Rochester Institute of Technology:

RIT Libraries Assignment Calculator: Project Management Tool for Students
<http://wally.rit.edu/researchguides/calculator/>

READING SCIENTIFIC PAPERS

If you are not used to looking at biological journals, you may find research papers hard to read and understand. Here are some guidelines to help you get the most out of a scientific paper.

■ Acquire some background knowledge.

Papers reporting original research are written for a relatively specialized scientific audience and thus assume some knowledge of the subject matter and the vocabulary. You can't, of course, become an expert in any field overnight. However, you can acquire some familiarity with the major ideas and key terms before tackling the primary literature. If you build this foundation, you will be less likely to feel overwhelmed later. See pages 6–7 for suggestions about locating general, introductory references. Also useful in this regard are scientific dictionaries and encyclopedias.

■ Read the Abstract first.

The Abstract—a brief summary of the paper—will give you an overview of the study and help you decide whether to read the rest of the paper. Don't feel intimidated by abstracts containing unfamiliar terms or ideas. Often the Introduction and Discussion sections will provide much of the explanatory material you will need for a fuller understanding. Your main task, at this point, is to decide whether the paper might be useful to you. Note that in some papers the main text is followed by a Summary section rather than preceded by an Abstract.

■ Understand the basic aims of the study.

If the paper seems relevant, read the Introduction carefully. *Why* did the author(s) conduct the research? What were the major hypotheses or predictions? Authors generally end the Introduction with a brief statement of their objectives; some also include a succinct statement of the major findings.

■ Don't get bogged down in the Materials and Methods.

It is easy to become overwhelmed by the specialized terminology found in many Materials and Methods sections. However, unless your research requires close examination of the methodology, you need not understand every procedural detail. Instead, work at grasping the author's basic approach. Try summarizing the methods in a few sentences, using your own words. Make sure you understand how the experimental design, use of controls, sampling techniques, or other methods relate to the specific objectives of the study. Once you have a sense of the major findings and conclusions, you can reread the Materials and Methods section, if necessary, to clarify selected points.

■ Read the Results carefully, focusing on the main points.

Do not panic if you don't understand the quantitative details. Focus first on the major qualitative findings and on the author's verbal summaries of quantitative data. These will lay the groundwork for understanding the more difficult material. The author may sum up main points in topic sentences (see p. 176) at the beginnings of paragraphs. Pay attention to comments in the text about figures and tables; they will help you pick out important quantitative trends or relationships. See Chapter 3 for a description of the common types of graphs used in biology and the sorts of information each is designed to convey.

■ Understand the author's argument.

Pay particular attention to the Discussion section, for it is here that the author attempts to tie together the various components of the study. Do the data support the author's conclusions, and if so, how? What does the author suggest is the major contribution of this study? What questions remain for further research? How can you use this paper in your own work?

■ Plan on rereading important papers, possibly several times.

Many of your sources may be too detailed or too difficult to understand in a single reading. Plan on returning to them once you have a firmer background and can better appreciate the specific contributions of each paper.

Also remember that when working with primary literature you can easily get sidetracked by the specifics of a particular study and thereby lose sight of the broader picture. Keep your own research in mind. Get a general grasp of each author's research, and then focus on whatever aspects are relevant to *your* objectives. The importance of a particular article may not be immediately apparent. You may need to skim through many papers at first to get your bearings and return later to those most central to your topic.

TAKING NOTES

■ Avoid plagiarism: take notes in your own words.

Plagiarism is the theft of someone else's words, work, or ideas. It includes such acts as (1) turning in a friend's paper and saying it is yours; (2) using another person's data or ideas without acknowledgment; (3) copying an author's exact words and putting them in your paper without quotation marks; and (4) using wording that is very similar to that of the original source but passing it off as entirely your own, even while acknowledging the source.

This last example of plagiarism is probably the most common one in student writing. Here is an example:

ORIGINAL PASSAGE	A very virulent isolate of *Alternaria mali,* the incitant of apple blotch, was found to produce two major host-specific toxins (HSTs) and five minor ones in liquid culture. The minor toxins were less active than the major ones, but were still specifically toxic to the plants which are susceptible to the pathogen. (Kohmoto and others 1976, p. 141)
PLAGIARIZED PASSAGE	Kohmoto and others (1976) found that a very virulent isolate of *Alternaria mali,* the incitant of apple blotch disease, produced two main host-specific toxins, as well as five minor ones in liquid culture. Although the minor toxins were less active than the major ones, they were still specifically toxic to the susceptible plants.

Although the writer has altered a few words here and there, the second passage is strikingly similar to the original. *It is still plagiarism if you use an author's key phrases or sentence structure in a way that implies they are your own, even if you cite the source.* The only way to make this passage "legal" as it now stands is to enclose everything retained from the original wording in quotation marks. Better yet, you should first determine which facts or ideas in a source are relevant for your purposes, and then put these in your own words and word order.

Plagiarism of this kind is usually unintentional, the result of poor note taking and an incomplete understanding of the ethics of research and writing. Typically the problem arises when you lean heavily on notes that consist of undigested passages copied or half-copied from the original source. These become the source of all the information and ideas for your paper. When you sit down to write the first draft, it is all too easy for this material to end up barely changed as the backbone of your paper. Thus your text becomes an amalgamation of other people's words disguised as your own. Even if you cite references for the facts and ideas, you are still guilty of plagiarism because the wording is not completely yours.

Another problem with this kind of note taking is that it reflects reading without thinking. It allows you to speed through a stack of references without necessarily understanding the material. It conflicts with your major purpose in conducting a literature review: to evaluate and interpret information on a subject. You need to start making judgments, comparisons, and contrasts while you are still working with the original sources; otherwise, your prose is just a mosaic of other people's material. Your own paper, like professional papers, should be more than just a sum of its parts.

Form the habit of taking notes mainly in your *own* words. If you are not used to doing this, you may be frustrated by the additional time it takes. However, once you start the first draft, these notes will save you much time and effort. You will have already worked through difficult material, weeded out many inconsistencies, responded to the conclusions of other authors, and made connections among related ideas. Much of the preliminary work will have already been done.

To take notes effectively you need to understand how to *paraphrase* and *summarize* material. A paraphrase expresses certain facts or ideas in different wording—your own—but in about the same number of words as the original. A summary expresses the important facts and ideas in fewer words than the original; for example, the Abstract of a research paper is a summary. Both paraphrasing and summarizing require that you understand material fully before you write about it. Although you will probably use both methods as you work through your sources, you'll find that learning how to identify and summarize the points that are *most* relevant to your particular needs is a highly valuable research skill. For example, the writer of the plagiarized passage above might have written the following in his or her notes, to be incorporated later in the final paper.

> *Kohmoto and others (1976) cultured the fungus Alternaria mali, which causes apple blotch, and isolated seven different toxins. Of these, two were particularly toxic to susceptible plants.*

■ Use an orderly system.

A common method is to use index cards, putting one idea or group of related ideas from a single source on each card. The cards thus contain manageable units of information and can be shuffled around at will as you organize your paper. However, such a method can be cumbersome, and many people feel constrained by the small size of the cards. Scientific topics often require longer, more detailed notes that cannot fit on index cards.

If you have a laptop computer, you may want to record notes directly into a computer file. Later you can cut and paste portions of your notes directly into your paper. An alternative method is to take notes on whole sheets of paper, writing on just one side so that you can cut, paste, and arrange notes later as you prepare the first draft.

Obviously, you need not take notes in complete sentences. In fact, if you try to restrict yourself to succinct phrases, you'll be even less likely to reproduce the exact wording of the original. If the author's own words *are* indispensable, enclose them within quotation marks along with the page number of the source. Do this for entire passages you wish to preserve, as well as for important words or phrases mixed in with your own notes:

> *J. concludes that "despite the predictive power and elegance" of the scientific method, it can give us only a "rough approximation" of what the natural world is like (Johnson 1933, p. 4).*

You also need a foolproof method to distinguish between an author's ideas and your own. For example, you might use a yellow marker to highlight your ideas or put your initials, the word *me,* or some distinctive symbol in front of any speculations and conclusions that are strictly your own:

> *B. suggests that light availability is the most important factor here. (me) What about moisture requirements? Not discussed.*

■ Be selective.

You will waste time and effort if you take copious notes on every source you encounter. Read first; take notes only when you have decided that the reference may be useful. Resist the temptation to photocopy *every* potentially useful source in its entirety; your research will be more productive if you try to make decisions, as you go along, about the relevance of each reference to your own objectives. If an article seems only peripherally connected or if you are not sure about its relevance, you may wish to photocopy the Abstract or other selected portions, but reserve your photocopying budget for those few papers that seem central to your work and that you may need to reread later.

■ Record full and accurate information about your sources.

Keep a master list of all the references you consult. Some people do this on whole sheets of paper or in a separate computer file; others list each source separately on index cards, so that when it is time to assemble the

Literature Cited (References) section, the cards can be shuffled and arranged in the proper order. Whatever system you use, be sure to separate the sources that you *cited* (referred to) in your paper from those you may have read but did not cite. Typically only cited sources are included in the Literature Cited section of biological papers.

Before you start your search, learn the kinds of bibliographic information you will need to report for each kind of reference you use (see Chapter 6). If you are not sure whether certain details are necessary, write them down anyway. It is easier to omit unneeded material when you eventually type your references than to spend time searching for missing publication dates or page numbers. If you photocopy part or all of an article, remember to write full bibliographic information about the source directly on the photocopy, so that you are never in doubt about its origin. Remember that Web sources need to be fully documented; see Chapter 6 for details.

CHAPTER **2**

Handling Data and Using Statistics

Whether in the laboratory or the field, much of your training as a biologist involves learning how to collect, manipulate, and interpret both *quantitative* (numerical) and *qualitative* (nonnumerical) data. The usefulness of these data depends on such factors as your experimental design or sampling procedures, your choice of equipment and skill in using it, the statistical techniques used for analysis, and (in a field study) the environmental conditions. You cannot foresee all the problems that might detract from the validity of your results. However, you can anticipate and avoid many potential difficulties by thinking carefully about the kinds of data you want to collect and how best to handle them. Thorough coverage of statistics and experimental design is well beyond the scope of this book; however, the general guidelines below may help you in the initial stages of your work. (For more specialized references, see Additional Readings on pp. 253–255.)

GETTING STARTED

■ **Start your research with questions leading to a specific prediction or hypothesis.**

This may sound like obvious advice to anyone acquainted with the scientific method (see Introduction). Nevertheless, beginners sometimes plunge headlong into research with little sense of purpose. Most experienced biologists do considerable thinking, planning, and preliminary data collection to get their bearings. Such work may be relatively unfocused and tentative, but it provides the grounding for the next step: the phrasing of key scientific questions and the formulation of specific predictions or hypotheses. Without *direction* and *purpose* in your research, you will simply amass large quantities of information but have little sense of what to do with it.

■ Understand how your specific study fits into a broader context.

Without such a theoretical background, you may waste time collecting inadequate data. If you are embarking on an independent research project, locate and read the established literature in your subject area (see Chapter 1). You should plan your own work so that it builds on and complements that of other researchers, rather than merely duplicating well-established findings. In addition, if someone has reported problems with a particular sampling method, experimental procedure, study species, or piece of equipment, you may be able to avoid the same difficulties.

Even in the case of structured lab work for a biology course, you need to appreciate *why* you are doing each activity. Such lab exercises are usually designed to illustrate general principles and to acquaint you with standard scientific procedures. Read your lab manual before coming to class, and try to understand how each experiment or field trip relates to a broader topic covered in lectures or the text. Only then can you collect and manipulate the right kinds of data effectively.

■ Identify the key variables so that you can plan your method of data analysis.

Are the important variables in your study *attributes* (qualitative categories such as smooth seeds versus wrinkled seeds, or red eyes versus white eyes)? If so, you will likely be interested in the *frequency* with which individuals fall into each category. Many studies involve *measurement variables* (quantitative characteristics such as weight, height, or temperature). Measurement variables can be *continuous* (taking any value within a given range, such as abdomen length in bumblebees) or *discontinuous* (taking only integer values, such as the number of leaves per zinnia plant). A third class of variables, *ranked variables,* includes ones that are ranked in order of magnitude, as in the case of the order of flowering (first, second, third, and so on; or 1, 2, 3, . . .) among a group of plants.

You will also need to determine what you will consider *independent* as opposed to *dependent* variables in your study. A dependent variable is so named because its value depends on (is affected by) the value of one or more independent variables. In other words, a dependent variable is a *function* of an independent one. For example, the growth rate of a bacterial culture is a function of temperature, transpiration rate in plants is a function of the relative humidity of the surrounding air, and so on.

The way in which you categorize and measure the key variables in your study will play a large role in determining the kinds of statistical analyses you'll need to use later.

■ Decide which variables you will hold constant or control for, either experimentally or statistically.

When investigating differences between two or more samples, or between expected and observed results, you must attempt to minimize the influence of some variables while isolating and evaluating the effects of other variables that are of interest to you. All this is part of careful experimental design—or in the case of descriptive field data, thoughtfully planned observation or sampling techniques.

Suppose, for example, you are studying the effect of pH on a particular enzyme-catalyzed reaction. Here the independent variable is pH; the dependent variable, the reaction rate, is expected to vary with pH conditions (and hence enzyme activity). You plan to manipulate values of pH in the laboratory, and to collect data on reaction rates under these different conditions. However, you assume that temperature will also influence the rate of the reaction; therefore, it will be necessary to *control* for temperature—for example, by conducting the experiment in a water bath under constant temperature conditions.

As a field example, consider a field study of mate-guarding by male dragonflies. You predict that males will spend less time perching and more time guarding their mates under certain conditions—for instance, when the population density is high. However, you suspect that ambient temperature may affect both population density and perching behavior. Although you can't control temperature conditions in the field, you can try to minimize the unwanted influence of temperature in your study by restricting your observations to a narrow, arbitrary range of temperature conditions, say 25–27°C.

As still another example, suppose you are testing the effectiveness of a 0.1% aqueous solution of gibberellic acid in stimulating radicle (embryonic root) elongation in germinating lupine seeds. You prepare two samples, each containing 24 seeds that have been soaked in distilled water for 96 hours. The seeds in one of these samples are treated with 10 mL of gibberellic acid solution, whereas the seeds in the second, *control* sample receive 10 mL of water only. You then subject both samples to the same temperature, light, and humidity conditions by conducting the experiment in a growth chamber. If, after an appropriate period of incubation, you find statistically significant differences in radicle length between the treatment and control groups, you can reasonably attribute these differences to the effect of gibberellic acid.

You can also *statistically* control for a particular "nuisance" variable, although this procedure should never substitute for careful use of experimental controls. For example, partial correlation analysis allows you to study the association between two dependent variables of interest while controlling for the effects of a third. Analysis of covariance is appropriate for examining the relationship between a dependent and an independent variable while controlling for the effects of a second independent variable. Check with your instructor for guidance on these and other experimental procedures. (See also the references in Additional Readings, pp. 253–255.)

■ Where appropriate, randomize your method of data collection.

Where individuals must be allocated to different treatments or must be selected from a larger group for measurement, you should avoid possible bias by choosing those individuals *randomly*. For example, suppose that in a plant growth study you have 200 bean plants growing in trays and wish to harvest 10 plants each week for dry weight determination. Choosing individuals randomly means that each plant in the tray has an equal chance of being selected; thus there is no danger of subconsciously picking plants of a particular size or appearance that might match the outcome you expect, or hope, to occur. One way to ensure randomization is to number each plant at the beginning of the experiment and then use a table of random numbers to select the 10 plants to be harvested each time. Such tables are found in statistical handbooks; also, many computer statistical applications (and even some calculators) have random number generators that make the task even easier.

RECORDING AND ORGANIZING YOUR FINDINGS

■ Devise effective data sheets.

Decide exactly what information you plan to record and in what form to record it. Make up a preliminary data sheet, using headings to create a series of columns in which to enter the values you think you will need.

Figure 2.1 (p. 37) is a sample data sheet for a laboratory project on the role of phytochrome in seed germination, using five different plant species. Its orderly arrangement allows you to record data quickly and systematically. Space is provided for recording the species, the date (useful for record-keeping purposes), and germination differences among the treatment groups. Two of the column headings already include the unit of measurement (here, percent). Unvarying data—for example, incubation condi-

Role of phytochrome in seed germination

Date _____

Plant species _____

No. of seeds per treatment _____

TREATMENT	% GERMINATION IN REPLICATE				MEAN % GERMINATION
	1	2	3	4	
Dark (control) R^1 FR^2 R→FR R→FR→R					

^{1}R = red light ^{2}FR = far-red light

Light intensity = _____ $\mu W\ cm^{-2}$

FIGURE 2.1 Sample data sheet for a laboratory study

tions and exposure times for red and far-red light—are recorded elsewhere in notes for the Materials and Methods section of the lab report.

For many researchers, spreadsheet software programs have greatly simplified the process of recording and analyzing data. Figure 2.2, created using Excel (Microsoft Corp.), is a computer-generated version of the data sheet in Figure 2.1. Other competing products, such as Corel Quattro Pro (Corel Corp.) and 1-2-3 (Lotus Development Corp.), work similarly to Excel. Note that, as in Figure 2.1, columns in Figure 2.2 are set aside for raw data pertaining to the different treatment groups. Additional columns (F and G) contain formulas set up in advance by the researcher for calculating particular descriptive statistics (here, means and standard deviations). As data are entered into columns B, C, D, and E, these statistics are automatically calculated and recorded. The accompanying graph is dynamically linked to the data table: the vertical bars are plotted from the values in column F, and the error bars from those in column G. The graph is updated

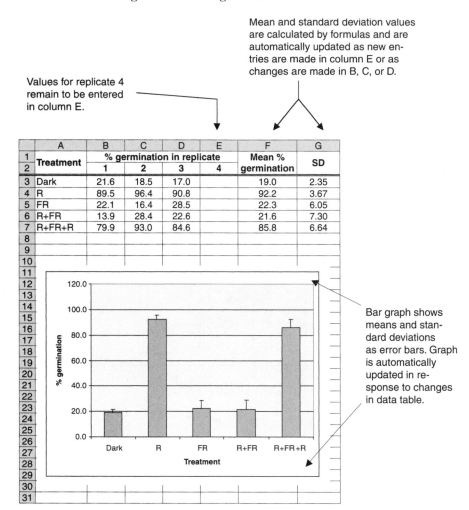

Mean and standard deviation values
are calculated by formulas and are
automatically updated as new entries are made in column E or as
changes are made in B, C, or D.

Values for replicate 4
remain to be entered
in column E.

	A	B	C	D	E	F	G
1	Treatment	% germination in replicate				Mean % germination	SD
2		1	2	3	4		
3	Dark	21.6	18.5	17.0		19.0	2.35
4	R	89.5	96.4	90.8		92.2	3.67
5	FR	22.1	16.4	28.5		22.3	6.05
6	R+FR	13.9	28.4	22.6		21.6	7.30
7	R+FR+R	79.9	93.0	84.6		85.8	6.64
8							
9							
10							

Bar graph shows
means and stan-
dard deviations
as error bars. Graph
is automatically
updated in re-
sponse to changes
in data table.

FIGURE 2.2   Computer-generated version of sample data sheet in Figure 2.1

each time data are entered; it can also be positioned anywhere on the spreadsheet and can be resized and edited at will.

Remember that a good data sheet is the product of much time and thought. Figuring out appropriate headings for all the columns forces you to evaluate the effectiveness of your methods and the rationale for collecting each set of information. However, even the most carefully planned data sheet may prove unsatisfactory the first time you use it, because methods or formats that seem feasible on paper do not always work in practice. You may need to modify your original sheet to account for problems you had not anticipated or for additional variables that need to be considered.

Here are a few more suggestions about recording data:

1. If you are recording data by hand, use pencil or waterproof ink, especially for fieldwork. After each laboratory or field session, look over the data sheets carefully. Check that your writing is legible. Record additional comments or questions immediately, before you forget them.

2. When transferring information to a computer data file, double-check your data for copy errors. Take the precaution of keeping backup copies of all files, and store these in a different place from the originals.

3. Keep your raw data sheets; you may need to look at them again later to check an anomalous result or to reexamine your data in light of subsequent research.

USING STATISTICS TO ANALYZE YOUR DATA

■ Make appropriate use of descriptive statistics.

A major use of statistics is to describe the data from a particular *sample* (for instance, 117 seed capsules collected from 63 *Linaria vulgaris* plants), generally with the aim of making broader statements about the *population* (here, the seed capsules of *all L. vulgaris* plants) from which the sample was drawn. Such descriptions include measures of the *variability* within samples. In fact, a major reason why scientists use statistics is because variability makes it difficult, if not impossible, to draw valid conclusions from data by mere inspection. Biologists must deal with three sources of variability: (a) measurement error, (b) environmental variation, and (c) genetic variability among the different individuals in a sample. To understand how measurement error can arise, suppose you set up 3 replicate test tubes and pipet 10 mL of water into each. It is highly unlikely that each tube would contain *exactly* 10.00 mL of water; rather, you might find that these tubes actually contained 9.95, 10.01, and 9.97 mL of water. Such lack of *precision* arises from every operation a researcher carries out, as well as from variation in the precision of instruments used.

Environmental variation can arise even in the most tightly controlled situations. For example, suppose you place the three tubes, above, in a rack in a "constant-temperature" water bath set to 37°C. If you could measure the temperature of the water circulating around each tube with a very accurate thermocouple, you would likely find that the temperature fluctuated around 37°C as the heating element switched on and off and as warmer and cooler circulating currents swirled around each tube. Moreover, you might well find that such variation was not identical for the three different tubes.

Biologists often describe samples using two different measures. One of these is the *mean* ($\overline{X}$), or arithmetic average of a set of values (that is, the sum of the values divided by the number of values). This measure is what we refer to when we talk about the "average."

The second commonly used descriptive measure in biology is the *standard deviation* (symbolized *s* and abbreviated SD), which gives a sense of the variability of the data within a sample. Two sets of data may have identical means but different standard deviations. For example, consider the following two sets of data: (a) 3, 9, 14, 19, 24, 29, 37, 41, and (b) 21, 21, 22, 22, 22, 22, 23, 23. Each data set consists of eight values; in both cases, the mean is 22.0, but even a cursory inspection shows that sample (a) has greater variability than sample (b). The procedure for calculating the standard deviation of sample (a) is shown below. First, the difference between the sample mean and each measurement value ("Deviation" column) is calculated. These differences are then squared ("Deviation squared" column) and summed, and the resulting sum is divided by the number of values minus one. The end result of these operations is a statistic called the *variance;* its square root is the standard deviation.

For the data above, the variance (s^2) =

$$\frac{\Sigma(X - \overline{X})^2}{n - 1} = \frac{1242}{8 - 1} = 177.42857,$$

and the standard deviation (*s*) is $\sqrt{177.42857} = 13.32023$. How should this value be reported? The accepted rule is to round derived statistics (mean, standard deviation, etc.) to the same level of precision as the individual measurements. Thus, if the measurements are made to two decimal places, the mean and standard deviation should also be reported to two decimal places. The exception is for measurements that are integers: in this

Sample (a) value (X)	Deviation ($X-\overline{X}$)	Deviation squared ($X-\overline{X}$)^2
3	-19	361
9	-13	169
14	-8	64
19	-3	9
24	2	4
29	7	49
37	15	225
41	19	361

Sum of squared deviations
$$\Sigma(X-\overline{X})^2 = 1242$$

case, it is customary to report the mean and standard deviation to one decimal place. Since the measurements above are integers, we would round both the mean and standard deviation to one decimal place (22.0 and 13.3, respectively).

Try the above procedure yourself using the data in sample (b); you should obtain a standard deviation of 0.8 (after rounding). Although both samples have the same mean, the variability of the first sample is much greater (SD = 13.3) than that of the second sample (SD = 0.8). Because mean values on their own convey only limited information, they are usually reported along with the sample size (number of measurements), as well as the standard deviation or some other measure of variability. Where consideration of the highest and lowest values is important, the *range* may be given instead of, or in addition to, the standard deviation. The range is defined as the maximum minus the minimum value. For sample (a) the range is 38, whereas that for sample (b) is only 2.

Certain measures derived from the standard deviation—for example, the *standard error* (symbolized $s_{\overline{X}}$ and abbreviated SE) and 95% *confidence intervals*—are often used to indicate the reliability of a sample mean as an estimator of the true mean of the population from which the sample was drawn. There is a 68.26% chance that the interval composed of $\overline{X} \pm s_{\overline{X}}$ includes the true mean of the population. A 95% confidence interval (derived mathematically from the standard error) is a wider interval with a 95% chance of including the true population mean. Thus, the larger the standard error (and hence the 95% confidence interval), the less reliable is a sample mean as an estimate of the population mean. See Figures 3.2, 3.4, and 3.10 in Chapter 3 for graphical depiction of standard errors.

Be sure you understand your purposes for reporting *any* descriptive statistics about a set of values. Remember that misuse of statistics can distort data rather than illuminate them. Consult your instructor or see the Additional Readings on pp. 253–255, for guidance in making appropriate use of these and other statistical measures.

■ When using statistical tests, understand the role of a null hypothesis.

A *null hypothesis*, or hypothesis of no difference, assumes that there is no difference among two or more populations from which samples have been drawn or no difference between results obtained and those expected. For example, your null hypothesis in an insect physiology experiment might be that there is no difference in the adult body weight of cockroaches raised on two different diets. In a lab exercise for a genetics course, your null hypothesis might be that the proportion of yellow seeds to green seeds in peas does not differ from the expected ratio of 3:1. Biologists subject data to particular statistical tests in order to determine if they can *reasonably reject* the null hypothesis in favor of an alternative hypothesis.

Suppose in the lupine study mentioned earlier (p. 35), you obtain the following measurements of radicle length in your two groups:

Radicle length (mm)

Treatment group (+ gibberellic acid)				Control group (no gibberellic acid)			
3.6	7.0	4.4	5.3	5.5	6.1	6.5	3.0
6.7	11.0	4.9	2.1	5.8	3.7	6.3	5.1
4.5	8.8	4.9	7.1	3.5	4.9	3.4	4.4
5.2	4.7	10.4	3.1	4.5	3.8	5.6	7.8
4.6	7.1	5.1	4.0	3.6	4.2	3.8	5.8
8.3	4.8	6.8	5.8	6.6	6.2	4.8	6.0

From these data, you calculate a mean radicle length of 5.8 mm (SD = 2.2 mm) in the seeds treated with gibberellic acid and a mean length of 5.0 mm (SD = 1.3 mm) in the untreated seeds. The null hypothesis assumes that treatment with gibberellic acid has no effect; that is, that mean radicle lengths do not differ between populations of treated and untreated lupine seeds. The alternative hypothesis is that a difference in mean radicle length does in fact exist. Because the mean lengths are numerically different, you might be tempted to conclude that the null hypothesis is false and that the two groups do differ in mean radicle length. You might further propose that this difference is tied to the presence or absence of gibberellic acid. However, there is still some possibility that such a finding might be due purely to chance and that the null hypothesis is true. This is especially likely if the variability in each sample is large. Doing the appropriate statistical test gives you a measure of the likelihood of wrongly rejecting the null hypothesis.

In this case, you can perform either a *t*-test or an analysis of variance (ANOVA). A *t*-test is appropriate only for a two-sample case; ANOVA is more generally applicable to comparisons of two or more samples. In the present, two-sample case, the two tests are equivalent. Each of these methods generates a test statistic that you compare with a critical value in a table to determine the probability (*P*) of getting a difference between the sample means as large as you observe (or larger) by chance, if the null hypothesis is, in fact, true. Biologists generally feel confident about rejecting the null hypothesis if this probability is less than 5% (expressed as *P*

< 0.05). Results associated with $P < 0.05$ are generally considered *statistically significant*. Of course, rejection of the null hypothesis does not *prove* it to be false; rather, it suggests that the probability of wrongly rejecting the null hypothesis, on the basis of your experimental data, is sufficiently small as to make this error unlikely. In the case of the lupine study (p. 35), a *t*-test would lead us to conclude that there is no significant difference in mean radicle length ($P > 0.05$) between treated and untreated seeds. That is, we cannot reject the null hypothesis.

■ Make sure your data meet the assumptions of the statistical tests you use.

Understand the rationale for using the particular analysis you choose for your data. Some statistical tests, such as chi-square, are used for frequency data. Many others, collectively called *parametric* tests, are designed for measurement variables. Some examples of parametric tests are analysis of variance, *t*-tests, least-squares regression analysis, and Pearson correlation. Do not automatically assume that a parametric test is appropriate for your analysis. All parametric tests make certain assumptions about the samples in relation to the populations from which they are drawn. Specifically, for each sample, these tests assume that quantities called *residuals* (each being a sample value minus the group mean) come from normally distributed populations of equal variance. (When a large number of values from a normally distributed population are plotted according to their frequency, we get the familiar bell-shaped curve, which is governed by a complex mathematical equation.) Before conducting any parametric analyses, you should check how well your data fit the above assumptions. Computer statistical applications, such as those listed below, include various tests that serve this purpose.

When the assumptions of parametric analyses are not met, you have two options. One is to transform the original values to logarithms, square roots, squares, or reciprocals and then carry out the parametric analysis on these transformed values. Just remember that the transformed values must meet the assumptions of parametric tests. The second option is to use one of a variety of nonparametric tests. Such tests do not assume normal population distributions. However, they do require, among other things, that the distributions have a similar shape. Originally devised for ranked variables, nonparametric tests can be used for measurement variables as well, but they are generally less powerful than parametric analyses if the assumptions of the latter are met. Examples of nonparametric tests include the Kruskal-Wallis test, Mann–Whitney U-test, and Spearman rank correlation. Further information on parametric and nonparametric tests can be found in any general statistics text.

■ Investigate the statistical software applications available at your institution.

Computer software applications have greatly simplified data analysis, making even highly complex statistical procedures accessible to novices. For example, JMP (SAS Institute, Inc.) is a user-friendly statistical application available for Windows, Macintosh, and Linux operating systems. A full-featured version, called JMP-IN (Duxbury/Thomson Learning), is available for around $80 to students and faculty of degree-granting colleges and universities. Commonly used statistical tests are also available in the "Analysis ToolPak" of Microsoft Excel. Professional-quality software applications that may be available at your institution include SYSTAT, SPSS (both produced by SPSS, Inc.), SAS (SAS Institute, Inc.), Minitab (Minitab, Inc.), and S-Plus (Mathsoft Engineering and Education, Inc.) Certain applications specialize in producing high-quality graphs suitable for publication. For example, most of the figures in Chapter 3 were prepared using SigmaPlot (SPSS, Inc.). Graphs of various types can also be quickly produced using Microsoft Excel; these may be adequate for a lab report, but they generally lack the quality required by a professional journal. Check with your instructor about his or her requirements.

It is well worth the time and effort to determine which statistical applications are available through the computer facilities of your institution. You might also consider purchasing some software of your own. Remember, however, that although it is tempting to plug your data into any one of a number of statistical tests, many of them may be inappropriate. The responsibility for selecting the correct test still rests with you.

CHAPTER **3**

Using Tables and Figures

The text of scientific papers is often supplemented with tables and figures (graphs, drawings, or photographs). Such materials can convey certain types of information much more effectively than words alone. A table can help you compare the results of a variety of chemical analyses. A graph can illustrate the effect of temperature on the growth of bean seedlings. A line drawing can depict an aggressive interaction between two fish, and a photograph can record important features of your study site.

It is not true, however, that tables and figures are essential in a scientific report. Biology students sometimes think that *all* data must be tabled or graphed to produce "professional" results. Even in published papers, unnecessary tables or figures sometimes find their way into print, wasting the reader's time and raising printing costs. Your credibility as a scientist will suffer if you use tables or figures just to impress the reader, when simple text will do as well or better.

A crucial task in scientific writing is deciding on an effective way to display whatever relationships, patterns, and trends are present in the data. You need a good sense of what quantitative data you are seeking and why. What statistical analyses are most appropriate to address the question you have asked? This concern should have high priority at the beginning of your research, not at the end when there is no time to compensate for mistakes, problems, or oversights.

A second and related concern is how to incorporate your quantitative data smoothly into the manuscript. Can your measurements and statistical analyses be summarized easily in the text, or is a table called for? Would a graph be more effective than a table? Rarely is there only one way to depict a given data set; however, one way may be superior to others. You need to review the scientific argument you are making and determine how to display the evidence most convincingly.

45

TABLES

■ **Use a table to present many numerical values or (occasionally) to summarize or emphasize verbal material.**

A large amount of quantitative information can be tedious and cumbersome to report in the text. If you put it in a table, the reader can take in everything at a glance, as well as compare one item with another (see Table 3.1).

Do *not* use a table when it is more important to show a pattern or trend in the data; consider a graph instead. Also, never use a table (or figure) when you could just as easily summarize the same material in the text. Look, for example, at Table 3.2. This table is unnecessary because it does not contain much information. Its contents can easily be put into words: "In Poolville, New York, I found *Plantago lanceolata* at 3 out of 5 weedlots sampled (Johnson Road, Lynes Street, and Bryant Avenue, but not at Redstart Lane or Henry Street)."

Similarly, Table 3.3 is unnecessary. The main message (that seeds neither imbibed water nor germinated following any of the treatments) could easily be stated in the text.

Table 3.4 has not progressed very far from the raw-data stage. The results will be more meaningful after some simple calculations, allowing us to replace the whole array of numbers with two succinct sentences in the text:

> After being visually isolated from other fish for 2 wk, each of 11 male *Betta splendens* was shown its reflection in a mirror and ob-

TABLE 3.1. Soil analyses of four farm fields near Malverne, Vermont

Site	Macroelement concentration (g per 1000 g dry soil)					
	Total N	P	K	Ca	Mg	S
Schwab farm	0.1	0.9	12.2	15.1	11.6	0.6
Toomey farm	0.8	0.7	6.6	8.3	5.4	0.5
Charles farm	2.6	1.1	17.1	15.9	13.1	1.1
Hendrick farm	2.1	0.7	18.3	26.8	13.6	0.8

TABLE 3.2. Occurrence of *Plantago lanceolata* at five
weedlots in Poolville, NY, August 15, 2005

Location	Present (+) or absent (−)
Johnson Road	+
Lynes Street	+
Redstart Lane	−
Bryant Avenue	+
Henry Street	−

served for 30 s. Collectively, the fish responded by approaching
their images (X = 3.0 times, SD = 2.1), biting the mirror (X =
1.4 times, SD = 1.0), and erecting their gill-covers for an average
of 15.5 *s* (SD = 4.4).

Tables need not be filled with numerical values. They may also be used
(sparingly and carefully) to summarize numerous points, to summarize a
review of the literature on a topic, to compare and contrast related items,
or to list examples or details that would be too tedious to spell out sentence by sentence in the text (see Table 3.5).

TABLE 3.3. Effect of germination inhibitors on
Phaseolus vulgaris seeds[a]

Inhibitor	Concn. (mol L^{-1})	Imbibi- tion time (days)	Germina- tion time (days)	% germi- nation
Actinomycin-D	0.05	0	0	0
Coumarin	0.023	0	0	0
Thiourea	0.015	0	0	0
2,4-dinitro- phenol	0.028	0	0	0

[a]80 seeds in each treatment. Seeds treated for 2
days at 40°C.

TABLE 3.4. Response of male fighting fish
(*Betta splendens*) to their image in a mirror[a]

Fish nr	Duration of gill-cover erection (s)	Nr of approaches	Nr of bites
1	20	6	3
2	13	5	2
3	10	2	0
4	7	1	1
5	15	1	1
6	16	5	2
7	19	3	2
8	20	1	0
9	14	0	0
10	21	5	2
11	16	4	2

[a]Prior to the experiment, fish had been visually isolated from one another for 2 wk. Observation period for each fish was 30 s.

■ Number tables consecutively, and make them understandable on their own.

Give each table a number in the order in which you refer to it in the text (Table 1, 2, and so on). Even if there is only one table (or figure) in the paper, assign it a number.

A table should be able to stand apart from the text and still make sense to the reader; therefore the *title* must adequately describe its contents. Never use vague, general titles like "Field data," "Test results," or "Feeding studies." Instead, be more specific and informative:

Average accumulations of uranium in lichens sampled at Collins Forest Preserve, Montana

Indices of fear in preexperimentally conditioned and nonconditioned mice

Responses of inexperienced adult Blue Jays to Monarch and Viceroy butterflies

TABLE 3.5. Selected functions of cytokinins in plants

Function	Plant studied	Reference
Production and proliferation of genetic tumors	*Nicotiana tabacum*	Ichikawa and Syono (1991)
Control of apical dominance	*Nicotiana tabacum*	Faiss et al. (1997)
Regulation of cell cycle	*Arabidopsis thaliana*	Soni et al. (1995)
Stimulation of cell division	*Nicotiana plumbaginifolia, Arabidopsis thaliana*	John (1998) Riou-Kamlichi et al. (1999)
Induction of adventitious shoots in plant callus cultures	*Arabidopsis thaliana*	Brandstatter and Kieber (1998)
Delay of leaf senescence	*Nicotiana tabacum*	Ori et al. (1999)
Delay of exit of cells from root meristem	*Arabidopsis thaliana*	Werner et al. (2003)
Increase in tiller density in turf grasses	*Agrostis stolonifer, Poa pratensis, Cynadon dactylon*	Ervin and Zhang (2003)
Promotion of chloroplast maturation	*Arabidopsis thaliana*	Chory et al. (1994)

(*continued*)

TABLE 3.5. Selected functions of cytokinins
in plants (*continued*)

Promotion of cell expansion in leaves and cotyledons	*Raphanus sativum*	Rayle et al. (1982)
Promotion of lateral bud development	*Arabidopsis thaliana, Nicotiana tabacum*	Medford et al. (1989)
Maintenance of meristematic activity in shoots and roots resulting in promotion of organogenesis	*Arabidopsis thaliana*	Nishimura et al. (2004)

The title is usually written as a sentence fragment at the top of the table. (A sentence fragment is a group of words that lacks a subject, a verb, or both, or that does not express a complete thought. Table titles usually lack a verb.) By convention, only the first word of the title is capitalized. If necessary, you may add one or more full sentences after the title to give further explanatory information:

> Relative abundance of 11 cyprinid species at five lakes in Nassau County, New York, in 2004 and 2005. Fish were captured by seining on July 7–9 each year.

Footnotes also allow you to add clarifying information to a table. Unlike material appearing after the title, footnotes are less obtrusive and distracting. You can use numerals, lowercase letters, or conventional symbols (such as *) to mark footnotes. Put these as superscripts (such as [a]) at the appropriate places, and list the footnotes in order at the bottom of the table. Do not use more than one kind of designation for footnotes. If those in Table 1 have numerals, for example, use numerals throughout all your tables.

■ Use a logical format.

Arrange similar elements so that they read *vertically,* not horizontally. You can see the rationale for this if you compare Table 3.6 with Table 3.7. They display the same data, but Table 3.7 is more logically set out and therefore easier to read.

TABLE 3.6. Growth rate of cell cultures and activity
of ornithine decarboxylase (ODC) and succinate
dehydrogenase (SDH) in *Pseudomonas aeruginosa*
in response to various carbon sources

	Carbon source		
	Glucose	Sucrose	Mannitol
Growth rate (generations/h)	0.93	0.21	0.47
Activity of ODC (μmol CO_2/h)	12.6	6.9	1.5
Activity of SDH (mmol fumarate/h)	137.7	19.3	50.9

Arrange the column headings in a logical order, from left to right, and list the data down each column in a logical sequence. List the data neatly under the center of each column heading, and make sure that all dashes or decimal points are aligned (see Table 3.7). Put a zero before each decimal with a value less than one—for example, 0.23, not .23.

TABLE 3.7. Growth rate of cell cultures and activity
of ornithine decarboxylase (ODC) and succinate
dehydrogenase (SDH) in *Pseudomonas aeruginosa*
in response to various carbon sources

Carbon source	Growth rate (generations/h)	Enzyme activity	
		ODC (μmol CO_2/h)	SDH (mmol fumarate/h)
Glucose	0.93	12.6	137.7
Sucrose	0.21	6.9	19.3
Mannitol	0.47	1.5	50.9

■ Make the contents concise.

A table should display selective, *important* information — not peripheral details, repetitive data, or values that are uniform and unvarying. Unnecessary data will just clutter the table and obscure more relevant data.

For example, if photoperiod, temperature, or other conditions were the same for all lettuce seedlings in your plant growth project, then these are the standard conditions, to be reported in a footnote, the table title, or the Methods section. These data do not belong in the columns of the table. Similarly, some information useful during the research may not be important enough to mention at all; for example, the reference numbers you gave to each of the mice you dissected or to a series of sampling sites. Thus, the fact that certain data take up space in your lab notebook does not automatically qualify them for a prominent position in a table.

Abbreviations are used freely in tables to save space and make the contents easier to grasp. In practice you can abbreviate many words in tables that you would not ordinarily abbreviate in the text; for example, temperature (temp), experiment (expt), and concentration (concn). Abbreviations likely to be unfamiliar to the reader (including ones you devise yourself) should be explained in footnotes. No table should contain so many abbreviations that deciphering them becomes an end in itself, demanding unlimited patience from the reader.

Do not put units of measurement after the individual values in a table. To avoid redundancy, put them in the title or the headings (see Table 3.8, for example). You can also group related column headings under larger headings, so that identical details do not have to be repeated for each column. Row headings can be grouped in a similar fashion.

Make all explanatory material in tables brief and concise. Do not excessively repeat what you have already written in Materials and Methods; instead, give just enough information to refresh the reader's memory and make the table understandable in its context. If the same abbreviations or procedures apply to more than one table, give them only once. In later tables, refer the reader to your earlier footnotes: "Abbreviations as in Table 1" or "Test conditions as described in Table 4."

■ Check tables for internal consistency and agreement with the rest of the paper.

A common fault of student papers is that some tables (or figures) do not make sense in the context of the whole paper. For example, something in one table may conflict with a statement in the text; the data in a table may seem at odds with the trend shown in a graph; or the values in a table may not make numerical sense. Be sure to carefully proofread your tables. Is each value copied accurately from your data sheets? Do all the items in a group add up to the totals shown in the table?

TABLE 3.8. Frequency of budding cells and activity of ß-galactosidase in three strains of *Saccharomyces cereviseae* mating type *a* in the presence or absence of α-factor. Data are based on 10 replicates in each treatment.

Yeast strain	Mean % cells budding		ß-galactosidase activity (Miller units)	
	+ α-factor	− α-factor	+ α-factor	− α-factor
Wild type	34.0	51.8	872.4	52.0
Mutant 1	51.1	52.6	40.1	31.5
Mutant 2	49.3	43.9	489.1	56.7

FIGURES

■ Use a graph to illustrate an important pattern, trend, or relationship.

A graph is generally superior to a table when you are more concerned about the "shape" of certain data, for example, how activity in a beetle species is related to ambient temperature or how the concentration of glucose in the bloodstream changes over several hours. Table 3.9 displays data that become more interesting when we put them in a *line graph* (see Fig. 3.1), because we get a clearer picture of the survivorship of individual plant leaves through the 98 days of observation.

Note in Figure 3.1 that the *independent* variable (time) is plotted on the X-axis, and the *dependent* variable (percent of leaves surviving at each sampling time) is plotted on the Y-axis. The dependent variable is so named because its value depends on (is affected by) the value of the independent variable. Another way to say this is that the dependent variable is a *function* of the independent one.

Generally, the axes of a graph meet. However, if one or more points fall on one or both axes, it is permissible to shift the axes slightly apart for clarity. This has been done in Figures 3.1 and 3.6 and Figures 3.8 through 3.15.

TABLE 3.9. Survivorship of individual leaves of narrow-leaved willow herb (*Epilobium leptophyllum*) at Warden Brook, NY, June 2–September 8, 2005

Time (days)	Percent surviving	Time (days)	Percent surviving
0	100.0	56	47.1
7	99.2	63	45.2
14	97.4	70	38.2
21	93.6	77	24.9
28	85.0	84	17.3
35	77.3	91	10.8
42	72.5	98	0
49	62.8		

Figures 3.2 through 3.5 illustrate other types of graphs that biologists commonly use. Figures 3.2 and 3.3 are two different kinds of *bar graph.* In both figures, the Y-axis depicts a measurement variable (germination success and number of flowers per plant, respectively); in Figure 3.2 the X-axis is also a measurement variable (temperature), whereas in Figure 3.3 the X-axis depicts qualitative categories, or attributes (plant species). Figure 3.4 is a *histogram,* or frequency distribution, showing the frequency with which observations fall into a series of numerical categories plotted on the X-axis. Figure 3.5, a *scatterplot,* illustrates the correlation, or strength of association, between two variables. In this case, the association is between leaf length and width in a particular plant. The correlation coefficient (r) of 0.95 indicates a strong positive association between these two variables. In other words, below- and above-average values of leaf width are associated with below- and above-average values of leaf length, respectively.

Generally, a graph is *not* necessary when trends or relationships are statistically nonsignificant (see p. 88) or when data are so sparse or repetitive that they can be easily incorporated into the text. Figure 3.6 is an example of a useless graph because the results it portrays can be summarized in a single sentence: "Growth rates of hyphae of *Saprolegnia* and *Achlya* remained constant at 2.38 and 0.92 μm per minute, respectively, before and after treatment with penicillin."

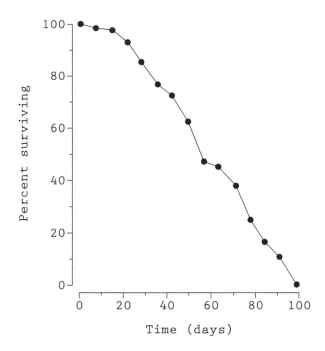

FIGURE 3.1. Survivorship of individual leaves of narrow-leaved willow herb (*Epilobium leptophyllum*) at Warden Brook, NY, June 2 - September 8, 2005.

■ Number graphs consecutively, separately from tables, and make them understandable on their own.

Each graph needs to be numbered in the order in which you discuss it. Use a separate series of numbers for graphs and tables—for example, Table 1, Table 2, Table 3; Figure 1, Figure 2, and so on. Use arabic numerals only, not roman numerals. Write out "Figure" in full in the *legend* (caption or title) of each graph. In the text, write "Figure" whenever it appears outside parentheses, but abbreviate it as "Fig." when you put it in parentheses. For example, "As shown in Figure 1, . . ." but "Germination rates (Fig. 1) . . ."

All figures should make sense apart from the text. The legend should be specific and informative, not vague as in "Weather conditions" or "Rainfall." Avoid simply repeating the labels of the two axes (for example, "Amount of rainfall vs time"). Instead, be more descriptive: "Rainfall fluctuations during the 1980 breeding season."

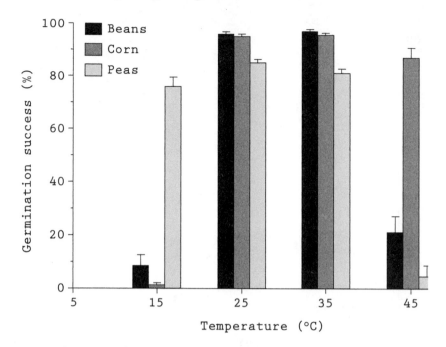

FIGURE 3.2. Germination of beans (*Phaseolus vul-garis*), corn (*Zea mays*), and peas (*Pisum sativum*) at four different temperatures. Vertical bars are +1 SE.

The legend is usually written as an incomplete sentence with only the first word capitalized. If necessary, you may follow it with a sentence or two of additional information, but be brief and concise (for example, see Figs. 3.6 and 3.10). Authors submitting manuscripts to journals typically type the figure legends on separate pages rather than on the figures themselves. For a student paper, you may wish to type or print each legend directly on the figure. (See pp. 201–204 for instructions on manuscript format.)

■ Plot data accurately, clearly, and economically.

Figures submitted to professional journals must meet high standards. A sloppily prepared graph will not show off the data to best advantage and may even cast suspicion on their quality. In student papers, too, accuracy and neatness count. Use standard graph paper, not lined notebook paper, to plot your data accurately. For a neat, professional-looking product, you can then trace this graph onto white bond paper. Also note that computer software applications have greatly simplified the statistical analysis and

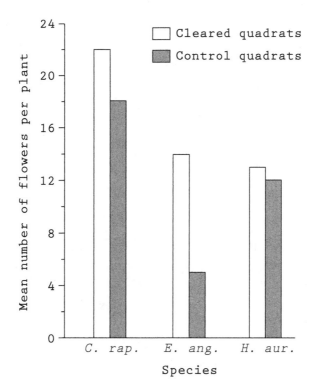

FIGURE 3.3. Production of flowers by three species of plants in the absence of interspecific competition (cleared quadrats) and under natural conditions (control quadrats). The plants were *Campanula rapunculoides, Epilobium angustifolium,* and *Hieracium aurantiacum.* Plotted are means for eight randomly chosen quadrats, each 1 × 1 m².

graphing of data (see p. 44). Check to see what is available at the computer center at your institution.

Remember that the independent variable belongs by convention on the X-axis. Label each axis clearly using large, easy-to-read lettering. Do not forget to specify the units of measurement. Numbers should also be large and legible. You need not number every major interval along an axis; instead, use index lines to indicate the halfway points between two numbers, as for example in Figures 3.1 and 3.5. This kind of shorthand prevents the graph from being cluttered with unnecessary numbers and gives you more space for essential information.

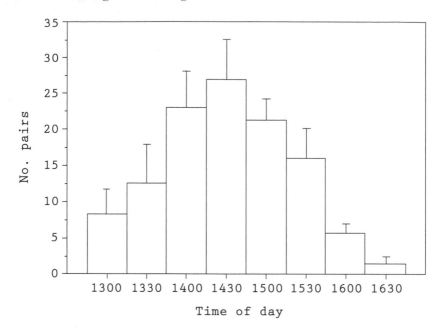

FIGURE 3.4. Number of ovipositing pairs of *Lestes con-gener* at sedge patches (total area = 17.9 m^2). Data are from a total of 93 censuses taken every 30 min on 12 different days, from 29 August - 18 September, 2005, at Ramsden Pond, Cazenovia, NY. Vertical bars are +1 SE.

Many graphs made by beginners waste large amounts of space. Look at Figure 3.7, for example. The Y-axis is shown originating at 0 even though the lowest measurement is 700 cells per mL; in addition, the X-axis is extended far more than is necessary given that the last measurement is taken at 60 h.

Figure 3.8 shows the same data in a more compact format. The Y-axis begins with the lowest value plotted, and the X-axis is only as long as needed. If you are used to seeing zero on every scale, this graph may look "wrong." However, indicating the zero mark should never be done at the expense of a graph's clarity and visual effectiveness.

Another way of saving space here is shown in Figure 3.9, where the Y-axis starts at zero but there is a break along it, indicating that unneeded values have been omitted.

You can also save space, as well as compare two or more data sets, by putting all of the data in a single figure, as in Figure 3.10. Be sure to label each data set clearly. Explain any symbols you use in the legend or in a key placed inside the graph itself. The latter method makes efficient use of

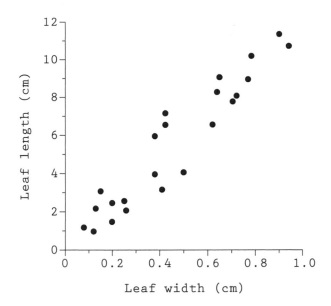

FIGURE 3.5. The relationship between leaf length and maximum leaf width in *Digitaria sanguinalis*. Data are from 23 randomly selected leaves from five plants, $r = 0.95$.

empty space. Do not pack *too* much information into a single figure, or the reader will be unable to make the necessary comparisons. A general rule of thumb is no more than four different symbols and no more than three different lines per graph. However, when many of the data points are close together, confusion and lack of clarity can quickly result, as shown in Figure 3.11. In this case, the author is better off preparing two separate graphs, one for mayflies and one for midges, and placing one below the other, so that the reader can easily make comparisons (see Fig. 3.12).

When you plot a series of sample means, it is customary to give the reader a sense of the variability of the data or the reliability of the sample means as estimators of the population means (see pp. 85–86). Figure 3.10, for example, shows the standard error (SE) for each mean as a vertical line extending up and down from each data point. These lines reflect the reliability of each mean as an estimator of the population mean. Using the same method to show the 95% confidence interval for each mean will convey similar information. If you wish to plot the variability of values within each sample, you could plot the standard deviation. If consideration of high and low values is important, you may prefer to show the range of values around each sample mean. Be sure to specify in the legend which statistics you have plotted.

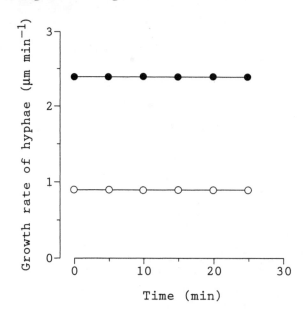

FIGURE 3.6. Growth rate of hyphae of the fungi *Sapro-legnia* (filled circles) and *Achlya* (open circles) at 20°C. Penicillin was added to the cultures at 10 min.

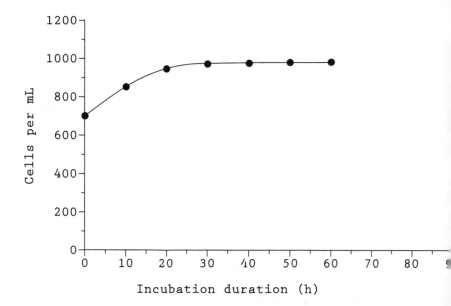

FIGURE 3.7. Change in population density of yeast cells growing at 27°C in malt extract broth.

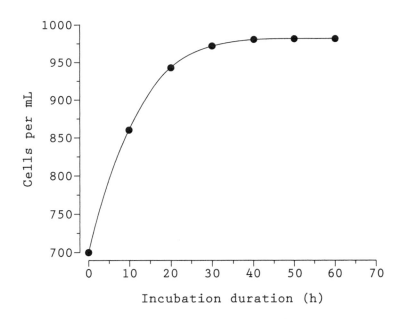

FIGURE 3.8. Change in population density of yeast cells growing at 27°C in malt extract broth.

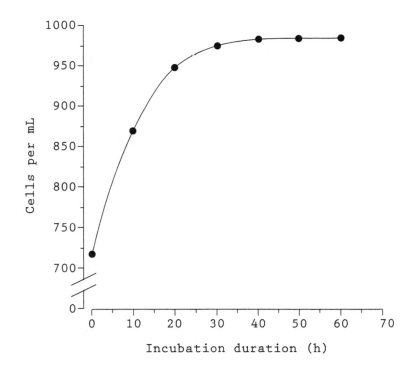

FIGURE 3.9. Change in population density of yeast cells growing at 27°C in malt extract broth.

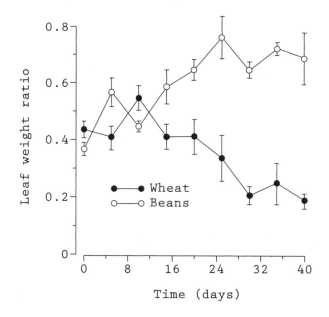

FIGURE 3.10. Growth rate of wheat (*Triticum vulgare*) and beans (*Phaseolus vulgaris*) as measured by the ratio of dry weight of leaves to dry weight of the whole plant. Ten plants of each species were sampled at each time. Vertical bars are ± 1 SE.

■ Depict data logically, in a manner consistent with your overall hypothesis.

Remember that the data you graph may be subject to more than one interpretation. You can bring out, hide, exaggerate, or misrepresent the message carried by a given set of data by the way you choose to draw the graph. For this reason it is vital that you understand your purpose in constructing the graph in the first place.

For example, does connecting all the dots on a graph always make sense? It depends on your data and how they were obtained and analyzed, as well as on your interpretation. In Figure 3.13 the data points are *not* connected; instead, the author has drawn a curve by hand to *approximate* the general downward trend suggested by the data. Most of the points are either touching or very close to this curve. Sampling error (virtually unavoidable in any study) accounts for the fact that all the points do not lie exactly on a smooth curve.

By contrast, in Figures 3.11 and 3.12, the author *has* joined all the points because his purpose is to show the high day-to-day variability in abundance of mayflies (or midges) at the study site. We assume here that

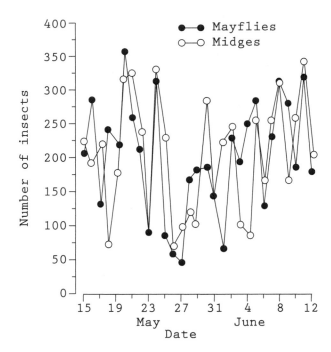

FIGURE 3.11. Abundance of mayflies (Ephemeroptera) and midges (Diptera: Chironomidae) at Hasior Pond, Mahoney, SC, in 2005. Data are from daily collections at 1700.

the big gaps between points are not due just to sampling variations; rather, they are meaningful fluctuations in themselves.

In Figure 3.14 the author has plotted a *statistically derived* (not hand-drawn) line of "best fit" (regression line) to the data points. Her purpose is to depict a specific mathematical relationship between two variables. The equation for the line, given in the legend, defines this relationship, and the R^2 value of 0.94 gives the additional information that approximately 94% of the variation in organic content (Y) is explained, or accounted for, by the line of best fit through the data points. Figure 3.15 illustrates a similar situation for a fitted curve. In both figures the authors have assumed that a cause-and-effect relationship exists between the two variables, that is, that values of the independent variable can be used to predict values of the dependent one. The scatter of the data points gives the reader a sense of the variability of the observations.

If you look back now to Figure 3.5 (p. 59), you will notice that the points are not accompanied by a fitted line. The author's purpose here is simply to show the interdependence of these two variables, not to suggest

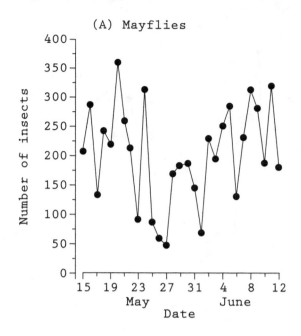

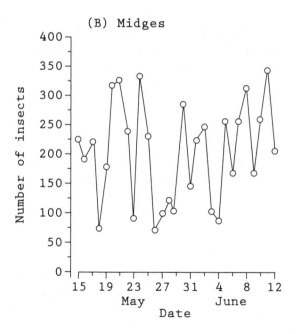

FIGURE 3.12. Abundance of mayflies (Ephemeroptera) and midges (Diptera: Chironomidae) at Hasior Pond, Mahoney, SC, in 2005. Data are from daily collections at 1700.

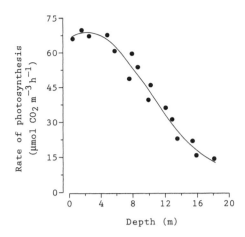

FIGURE 3.13. Rate of photosynthesis by marine phytoplankton as a function of depth. Data were collected
at Hilton Bay, SC, at 1100, July 31, 2000.

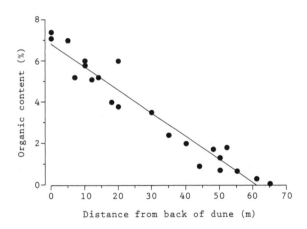

FIGURE 3.14. Decrease in organic content across the
width of a sand dune at Bradley Beach, Hilton Head
Island, SC. Determinations were made along a transect
placed perpendicular to the shoreline from the back of
the dune to the seaward side. Regression equation is
$Y = 6.83 - 0.11X$, $R^2 = 0.94$.

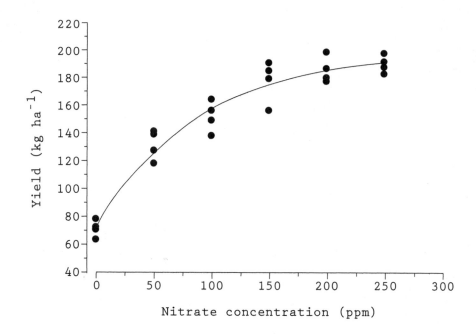

FIGURE 3.15. Yield of corn (*Zea mays*) in response to additions of nitrate fertilizer. Regression equation is $Y = 198.6 - 126.1e^{-0.012X}$, $R^2 = 0.95$.

that the values of one are controlled or predicted by the values of the other. As in other correlation studies (see p. 88), either variable could have been plotted on either axis, since no assumptions are being made about which one is dependent and which is independent.

Thus, constructing a graph is more than just an exercise in plotting points on ruled paper. It requires that you understand the rationale behind your quantitative methods. How is your interpretation of the data constrained by the statistical analyses you used? What assumptions are you making by describing your data in this particular way? Are these assumptions valid? Discussion of quantitative methods is beyond our scope here, but the subject is treated at length in the references listed at the end of this book.

CHECKLIST FOR TABLES AND FIGURES

- Is every table and figure absolutely necessary?
- Does every table have a title, every figure a legend?

- Do you use a figure rather than a table to show a pattern, relationship, or trend?
- Are data in figures, tables, and text consistent with one another?
- Are figures and tables numbered consecutively in separate series?
- Do you refer to every figure and table, and have you numbered them in the order in which they are mentioned in the text?
- Are tables and figures understandable apart from the text?

CHAPTER 4

Writing Lab Reports and Research Papers

A research paper is a report of original findings organized into several sections according to a format that reflects the logic of a scientific argument. The paper bears a specific, informative title, usually followed by a one-paragraph Abstract, or summary. In the body of the paper, the author first states the purpose of the investigation, placing the work in a broader scientific context (Introduction). Then the procedure is described (Materials and Methods), and the findings are presented (Results) and interpreted (Discussion). Finally, the author thanks others for their help (Acknowledgments) and lists the references that were cited in the paper (Literature Cited or References). Biologists follow this format closely when they write a paper for publication. As a biology student, you will use the same organizational method for research projects and laboratory reports.

Many writers feel that the fairly rigid structure of research papers and lab reports makes them relatively easy to write. Once you understand what goes where, you can break up the writing into discrete tasks and tackle each one in turn. In fact, you do not have to compose the paper in any particular order. Begin with the section that seems easiest; for many people this is the Materials and Methods. The Discussion and Abstract are usually easiest to write last. The Literature Cited section can be assembled as you go along, but you'll need to leave time at the very end to check it for completeness and accuracy.

Of course, the bulk of your writing will have to wait until you have analyzed all the results. However, if you try to visualize each step you take as part of a particular section of the paper, you'll be better prepared to start writing once the time comes. Moreover, you will stay in touch with the aims and progress of your study and be able to modify your approach, if necessary, while there is still time to collect additional data.

One good organizational method is to have a separate file folder, computer file, or notebook section for each section of the paper. While the study is still in progress, jot down procedural details, notes on the strengths and weaknesses of particular experiments, and speculations about the relevance of your findings. Record ideas, questions, and problems as they come to you, and file everything in the appropriate place. Do not rely on memory. As good as yours may be, you are bound to forget something unless you write it down. The act of writing will impose order and logic on your activities, especially at the start of the project when you may be inefficient and disorganized. Writing will also help you crystallize your thoughts, gain perspective, and perhaps see new patterns in the data. You may find that a casual thought captured on paper eventually proves to be a flash of insight.

TITLE

The title identifies the important contents of the paper and orients the reader by specifying the writer's major findings or perspective. For this reason, abstracting and indexing services rely heavily on the titles of papers to organize large quantities of scientific information. Researchers also scan titles to pin down the most relevant references. A vague or inaccurate title can waste a reader's time by suggesting, erroneously, that the paper contains certain information. Even worse, a good paper burdened with a bad title may never catch the eye or the interest of many of its intended readers.

Many writers compose the title last, once they have written at least one draft of the paper and have a better understanding of their purpose and scope. Others devise a working title *before* they do much writing, to help them focus their ideas; then they revise it later, if necessary. See page 202 for instructions on preparing the title page of the final manuscript.

■ Make the title informative and specific.

Organize your title around the important words in the study. Imagine that your paper appears in an index or abstract. How easily could it be located by someone looking up keywords pertaining to your subject area? (Biologists submitting papers to journals for publication are often asked to suggest several keywords pertaining to their manuscript.)

Your title should indicate the major focus of your study. The following titles, for example, are vague and uninformative compared with the revised versions following them.

VAGUE	Ecological Studies of Some Northern Lakes
SPECIFIC	Seasonal Algal Succession and Cultural Eutrophication in Three North Temperate Lakes

VAGUE	Dominance Behavior in Rhesus Monkeys
SPECIFIC	Androgen-Induced Social Dominance in Infant Female Rhesus Monkeys
VAGUE	Aspects of Territoriality in Lizards
SPECIFIC	The Adaptive Significance of Territoriality in Iguanid Lizards
VAGUE	Some Changes during Menopause
SPECIFIC	Early Menopausal Changes in Bone Mass and Sex Steroids
VAGUE	A Look at Fungal Toxins
SPECIFIC	The Role of Fungal Toxins in Plant Disease

■ Be concise.

Make every word count. Omit unneeded "empty" words at the beginning of the title.

WORDY	Preliminary Studies on Primary Productivity and Phytoplankton Diversity in Four Wisconsin Lakes
CONCISE	Primary Productivity and . . .
WORDY	Notes on the Effect of Ambient Temperature and Metabolic Rate on the Activity of New Mexican Lizards
CONCISE	Effect of Ambient . . .

Replace wordy phrases with shorter phrases or single words.

WORDY	Why Felids Copulate So Much: One Possible Model for the Evolution of Copulation Frequency in the Felidae
CONCISE	An Evolutionary Model for High Copulation Frequency in the Felidae
WORDY	Studies on the Reproductive Biology of *Drosophila,* Including Sperm Transfer, Sperm Storage, and Sperm Utilization
CONCISE	Sperm Transfer, Storage, and Utilization in *Drosophila*

■ Include appropriate taxonomic information.

If your work features a particular species or larger taxonomic group, specify this clearly in the title. Species should be described by their full Latin names—both genus and specific epithet. Include other information, if necessary, to orient the reader. Such information may include common names (if they exist) or the name of the family, order, or other important group to which the species belongs. Note how the following titles are more informative in the revised versions.

VAGUE	Mating Frequency in Butterflies
VAGUE	Mating Frequency in *Papilio*
SPECIFIC	Mating Frequency in Butterflies of the Genus *Papilio*

The first title is too general, unless the writer is discussing *all* butterflies; the second will be confusing to readers unfamiliar with insect (and especially butterfly) taxonomy.

VAGUE	Effect of Hormones and Vitamin B on Gametophyte Development in a Moss
SPECIFIC	Effect of Hormones and Vitamin B on Gametophyte Development in the Moss *Pylaisiella selwyni*

The first title conveys the purpose of the study, but we are left wondering *which* moss is being studied.

VAGUE	Pollination, Predation, and Seed Set in *Linaria vulgaris*
SPECIFIC	Pollination, Predation, and Seed Set in *Linaria vulgaris* (Scrophulariaceae)

Even botanically inclined readers may not be familiar with this particular species. Including the family (Scrophulariaceae) after the scientific name provides the useful information that this plant is in the snapdragon family.

VAGUE	Paternity Assurance in the Mating Behavior of *Abedus herberti*
SPECIFIC	Paternity Assurance in the Mating Behavior of a Giant Water Bug, *Abedus herberti* (Heteroptera: Belostomatidae)

As in the preceding example, the first title of this pair is not informative enough for most readers. Is this organism a mammal? A fish? A bird? The revised version tells us that it is an insect belonging to the order Heteroptera and family Belostomatidae.

Note that there are some species or larger groups whose scientific names do *not* need to be specified in the title, although it is not incorrect to do so. These organisms include those whose common names are familiar, well established, and even standardized (such as American birds), and organisms that are well studied or familiar to most scientists, such as cats or rats. However, it is still necessary to include precise taxonomic information in the text of the paper.

Finally, some species—for example, *E. coli* and *Drosophila*—are well known by their *scientific* names; they can appear without further explanation in the title.

■ Avoid specialized terminology, coined words, and most abbreviations.

As in a published paper, your title should be meaningful on its own; otherwise it will confuse or discourage potential readers. Write the title with your audience in mind. Use only those terms likely to be familiar to most readers—for example, your instructor and your classmates. Avoid esoteric or highly technical language, and do not use abbreviations except for widely recognized ones (such as DNA, RNA, or ATP).

ABSTRACT

■ Summarize the major points of the paper.

The Abstract is a short passage (usually 250 words or less) that appears just after the title and author(s) and summarizes the major elements of the paper: objectives, methods, results, and conclusions. It is typically written as a single paragraph. In the case of published works, a good Abstract helps researchers assess the relevance of a paper to their own research and thus decide whether or not they should read it completely. Scanning the Abstracts of papers is also one way that scientists, when pressed for time, keep abreast of recent literature.

Although readers see the Abstract first, it is easiest to write it last, once you have a good overview of the paper. One way to write an Abstract is to list, one by one, all the important points covered in each section of the paper. Write complete sentences if you can. Make successive revisions, paring the list down bit by bit, omitting peripheral topics and details, until you have revealed the "skeleton" of the study.

■ Be specific and concise.

As a summary, an Abstract must be both informative and brief. Avoid general, descriptive statements that merely hint at your results or serve only as a rough table of contents. Consider every sentence, every word. Could you say the same thing more economically?

The following Abstract from a published paper illustrates how the essential points of a study can be summarized in a single concise paragraph:

> The action of melanin-concentrating hormone (MCH) on melanophores was studied in 27 teleost species. MCH caused melanosome aggregation in all teleosts studied, including two siluroid catfish in which melanin-aggregating nerves are known

to be cholinergic. In most fish, the minimal effective concentration of MCH was estimated to be 10 μM, while in three swellfish examined, it was higher than 10 μM. The mode of action of the peptide was identical in either adrenergically or cholinergically innervated melanophores. It may act through specific receptors on the melanophore membrane. These results suggest that MCH may be a biologically active hormone common to teleosts.

(Nagal and others 1986, p. 360)

The Abstract below is too general: the student writer has left out important information and refers only vaguely to what was done, how, and why.

> This report is a study of the algae in Lebanon
> Reservoir, using a variety of collecting tech-
> niques. Over the study period, there was a
> gradual replacement of Cyanophyceae by Bacillar-
> iophyceae as a result of certain factors, such
> as temperature. Planktonic communities differed
> from attached communities because of morphologi-
> cal and physiological characteristics of the
> algal species in each community. There were
> marked differences in algal composition between
> the epilimnion and hypolimnion.

This Abstract could be improved if the writer used more specific language. As it stands, we are left with unanswered questions. What kinds of collecting techniques were used? Were there other important factors besides temperature? What "morphological and physiological characteristics" did the writer study? And so on. We also do not know where Lebanon Reservoir is or why the author did the study in the first place.

The Abstract below is unnecessarily wordy:

> This paper reports studies on the territorial
> behavior patterns shown by males of the dragon-
> fly species *Plathemis lydia* at a small pond in
> the vicinity of Earlville, New York. A total of
> 51 male dragonflies were marked with small spots
> of enamel paint applied to the abdominal portion
> of the body and were observed under natural con-
> ditions during the month of June 2005. It was
> found that males showed a strong tendency to

```
defend individual areas at the pond. These males
chased away other males of their species, and
also, on occasion, even males of other species
of dragonflies. Threat behavior, as opposed to
behavior involving physical contact, was the
most common aggressive behavior displayed by
males defending their areas against intruders.
Males typically remained at the same area of the
pond for two hours or longer on a single day.
Also, they usually were observed returning to
the same area on successive days. These observa-
tions suggest the possibility that male terri-
torial behavior in this species serves the
adaptive function of enabling the owners to mo-
nopolize particular areas. These areas may be
visited by females who are in the act of seeking
mates.
```

This Abstract used 192 words to describe what the same Abstract, written more concisely, can express in 104 words:

```
I studied the territorial behavior of male dragon-
flies (Plathemis lydia) at a small pond near
Earlville, New York. Fifty-one males were marked
with enamel paint on the abdomen and observed
under natural conditions during June 2005. Males
defended individual areas from male conspecifics
and occasionally from males of other species.
Aggressive interactions generally involved
threat behavior rather than physical contact.
Territory owners typically remained at the same
area for at least two hours and returned to the
same location on successive days. Territorial
behavior in this species may be adaptive for
males, enabling them to monopolize areas likely
to be visited by females seeking mates.
```

In the revised version, the author has replaced long, wordy constructions with shorter, more economical wording: for example, "the territorial behavior patterns shown by males of the dragonfly species *Plathemis lydia*" has become "the territorial behavior of male dragonflies (*Plathemis lydia*)." Many unnecessary words ("a total of," "it was found that," "these observations suggest the possibility that") have been omitted. The resulting Abstract conveys the same information much more effectively. (See pp. 187–189 for a more complete discussion of wordiness.)

The Abstract below is too long because the author included more information than was necessary. What parts should be left out?

The algal succession occurring with autumn overturn was studied in the phytoplankton community
of a hard water pond near Madison, New Jersey.
The pond can be sampled easily from all sides
and offers excellent opportunities for ecological research. This study was done as a field
project for Biology 314. The dominant species
in October 2005 were *Dinobryon sertularia*,
Scenedesmus quadricauda, *Cryptomonas erosa*, and
Cyclotella michiganiana. The first three species
showed significant declines in December, after
fall overturn, whereas the last species became
much more prominent.

We do not need to know that the pond is a good place to do research or that the study was done for a particular biology course.

■ Make sure the Abstract can stand on its own and still make sense to the reader.

Like the title, the Abstract of a published work will be read by many people who may not read the text of the paper. Therefore, the Abstract must be independent: readers should be able to understand it without being familiar with the details of the study. Omit abbreviations (except for widely known ones such as DNA), and use only those technical terms likely to be familiar to your audience. If certain abbreviations or specialized terms are absolutely necessary, then explain them. Do not refer to materials that may be inaccessible to your readers, such as your tables and figures, and avoid references to other literature, if possible.

The following Abstract, taken from a student research report, is difficult to understand without having first read the paper:

This paper describes a behavioral and computer
simulation study of cognitive differences between five-year-old boys and girls presented
with tasks of either Group I, II, III, or IV
difficulty level. We followed a modified version
of the method developed by Wolf (1955, 1977).
When compared with boys, girls scored higher on
what we designated "prepared tests," whereas
boys scored higher on "stimulus-response tests"
(Fig. 3). We concluded that . . .

The author of this Abstract provides little information we can understand without having first read the paper. We have no way of knowing what "Group I, II, III, or IV" stands for, what Wolf's method was, what the author's definitions of "prepared" and "stimulus-response tests" were, or what Figure 3 is all about.

Now look again at the Abstract by Nagal and others (1986) (pp. 72–73). Notice that the abbreviation MCH is explained and the language is straightforward. The few technical terms that are used (for example, "melanophores" and "cholinergic") are likely to be familiar to most readers in this field.

INTRODUCTION

The Introduction of a research paper sets the stage for your scientific argument. It places your work in a broad theoretical context and gives readers enough information to appreciate your objectives. A good Introduction "hooks" its readers, interesting them in the study and its potential significance. Thus you, as a writer, must have a firm grasp of the aims, principal findings, and relevance of your research. You may find that the Introduction is easiest to write *after* you have drafted the Materials and Methods, Results, and Discussion sections and have a clearer understanding of just what you are introducing.

■ Orient the reader by summarizing pertinent literature in your field.

An effective way to organize the Introduction is to proceed from the general to the specific, starting with a review of current knowledge about the topic and narrowing down to your specific research problem. Introduce key concepts, define specialized terms, and explain important hypotheses or controversies. As you do so, *document* your writing by citing key references in the field — more general ones first, followed by studies closest to your own research. (See Chapter 6 for instruction on how to document sources.) In this way you can sketch out a framework for the study, orient your readers, and prepare them for what is to follow.

Do not make the Introduction *too* broad or *too* detailed. This is not the place to show off your knowledge about the subject, list every available reference, or repeat material found in any elementary text. Most published papers have short Introductions (often only a few paragraphs) because the writer is addressing readers with backgrounds similar to his or her own. It wastes journal space and the reader's time to give an exhaustive literature review. Similarly, in a paper for a course, write for your classmates and your instructor—people with at least a beginning knowledge of the subject. Discuss only the most relevant concepts and references, and get quickly to the point of the paper.

■ Explain the rationale for the study and your major objectives.

After explaining the broad theoretical context, you are ready to clarify the special contribution your own study makes. How does your work fit in with that of other researchers? What special problem does your study address? What *new* information have you tried to acquire? Why? How? In other words, why (apart from course requirements) are you writing the paper in the first place? Most authors end the Introduction by stating the *purpose* of the study. For example:

> The purpose of this study was to describe the dominant fungi associated with decomposing leaf litter in a small woodland stream.

> In this paper, I report laboratory observations on the effects of crowding on the behavior of juvenile Atlantic salmon (*Salmo salar*).

Remember to clarify the purpose of your paper by first explaining the rationale for the study. What needs to be improved in the following Introduction?

```
The purpose of this project was to study the
algae at two different sites in Payne Creek,
near Stillson, Florida. Chemical and physical
parameters were also considered. The study in-
volved algal distribution in several microhabi-
tats of both stream and marsh environments. This
project was part of a larger class study for
Biology 341.
```

A major problem here is that the writer has not linked the study to a broader conceptual framework. She begins abruptly with a vague statement of her objectives, with no explanation of why they are important. The passage also contains irrelevant information: it is not necessary to state that the study was part of the requirements for a particular course.

By contrast, the writer of this Introduction gives the reader a clear idea of what the objectives are and why they are important:

> It is well known that males of many species of dragonflies (order Odonata) guard their mates after copulation while the eggs are being laid. In some species the male hovers over the female and chases away other males; in others, the male remains physically attached to the female as she moves around the breeding site. Many authors have discussed the adaptive significance of mate-guarding, particularly with respect to ways in which it may increase the reproductive success of the guarder (for example, Alcock 1979, 1982; Sherman 1983; Waage 1984). A more complete understanding of the evolution of this behavior is dependent on detailed studies of many odonate species. In this paper, I describe mate-guarding in *Sympetrum rubicundulum* and discuss ways in which guarding may be adaptive to males.

Some authors end the Introduction by summarizing their major *results* in a sentence or two. This tactic gives readers a preview of the major findings and may better prepare them for the scientific argument that follows. Other writers, along with some journal editors, criticize this practice, arguing that results are already covered in their own section and in the Discussion and Abstract. Ask your instructor what he or she prefers.

MATERIALS AND METHODS

■ Include enough information so that your study could be repeated.

Your methodology provides the context for evaluating the data. How you made your measurements, what controls you used, what variables you did and did not consider—all these things are important in molding your interpretation of the results. The credibility of your scientific argument depends, in part, on how clearly and precisely you have outlined and justified your procedures.

Furthermore, one of the strengths of the scientific method is that results should be reproducible using similar materials and methods. It is not unheard of for a scientist to repeat someone else's experiment and get different results. These conflicting data then point to factors that may have been overlooked, perhaps suggesting different interpretations of the data.

Finally, a complete and detailed Methods section can be enormously helpful to others working in the same field who may need to use similar procedures to address their own scientific problems.

What kinds of information should you include? See the guidelines below.

Materials

1. Give complete taxonomic information about the organisms you used: genus and specific epithet as well as subspecies, strains, and so on, if necessary. Specify how the organisms were obtained and include other information pertinent to the study, such as age, sex, size, physiological state, or rearing conditions.

> *Cladosporium fulvum* race 4 was obtained from Dr. George Watson, Department of Biology, Colgate University, Hamilton, New York. Stocks were maintained in sterile soil at 4°C and were increased on V8 juice agar at 25°C in the dark. . . .

> Adult American chameleons (*Anolis carolinensis*), purchased from a local supplier, were used for all experiments. They were kept in individual terraria (30 × 30 × 30 cm³) for at least seven days prior to the start of any study. They were provided with a constant supply of water and were fed crickets, mealworms, and other insects every two days. All chameleons were exposed to 15 h of fluorescent light daily (0800–2300), and air temperatures were kept at 30°C. . . .

2. If you used human subjects, give their age, sex, or other pertinent characteristics. Biologists submitting papers for publication may need to demonstrate that subjects have consented to be involved in the study.

3. Describe your apparatus, tools, sampling devices, growth chambers, animal cages, or other equipment. Avoid brand names, unless necessary. If some materials are hard to obtain, specify where you purchased them.

4. Specify the composition, source, and quantities of chemical substances, growth media, test solutions, and so on. Because they are more widely understood, use generic rather than brand names.

> Sodium citrate, sodium pyruvate, and hydroxylamine were obtained from Sigma Chemical Company, St. Louis, Missouri. All chemicals were of reagent grade. . . .

For the first series of electron microscopy studies, tissue samples were fixed in 2.8% ultrapure glutaraldehyde, 0.57 mol L^{-1} glucose, and 0.10 mol L^{-1} sodium cacodylate. . . .

5. If detailed information about any of the materials is available in a standard journal, then avoid repetition by referring the reader to this source.

I used an intermittent water delivery system similar to that described by Lewiston (1950). . . .

Tryptone–yeast extract broth (Pfau 1960) was used to cultivate bacterial strains. . . .

Methods

1. Describe the procedures in detail. Do not forget crucial details such as temperature conditions, pH, photoperiod, duration of observation periods, sampling dates, and arbitrary criteria used to make particular assessments or measurements. If you used a method that has already been described in a standard journal, you need not repeat all this information in your own paper; just cite the reference.

Mycelia were prepared using the fixation and embedding procedures described by Khandjian and Turner (1971). . . .

However, if the reference is hard to obtain (for example, *The Barnes County Science Newsletter*), or if you altered someone else's methods, then supply full information about your procedures.

2. For field studies, specify where and when the work was carried out. Describe features of the study site relevant to your research and include maps, drawings, or photographs where necessary. If published information already exists on the area, cite sources.

This study was conducted during June and July 2005 at Bog Pond, 3 km northwest of Barrow, West Virginia. The general habitat of this pond has been described elsewhere (Needham 1967; Scott 1981). The pond is permanent and contains floating and emergent vegetation (mostly sedges, rushes, and algae). It has an area of approximately 1.5 ha. . . .

3. Commonly used statistical methods generally need no explanation or citation; just state for what purpose you used them. If you used less familiar or more involved procedures, cite references explaining them in detail and give enough information to make your data meaningful to the reader.

■ Organize your material logically.

Because the Materials and Methods section contains so many important details, it is easy to forget some, particularly because you are so familiar with the subject. It is also easy to let this section become a confusing, rambling conglomeration of details, with little unity or coherence. Organize this section carefully using an outline, a list, or a plan, along with the detailed notes you compiled while doing the research.

A typical approach is to begin with a description of important materials (study species, cell cultures, and so on) and to move on to the methods used to collect and analyze the data. Field studies often start with a description of the study site. Describe your procedures in a logical order, one that corresponds as closely as possible to the order in which you discuss your results. You may also group related methods together.

Remember that the Materials and Methods section is still part of your text and must be readable. Do not let your paragraphs become disorganized collections of choppy sentences, as in the example below:

FAULTY   Golden hamsters (*Mesocricetus auratus*) used for this research were adult males. Temperature conditions were kept at 22–24°C. Animals were fed Purina chow. Hormonal studies were performed on 23 individuals. The photoperiod was 16 h. Animals were housed with littermates of the same sex, and feeding was once each day. All hamsters had been weaned at three weeks.

REVISED   Hormonal studies were performed on 23 adult male golden hamsters (*Mesocricetus auratus*). All had been weaned at three weeks and housed with littermates of the same sex. They were reared under conditions of 22–24°C and a photoperiod of 16 h and were fed Purina chow once daily.

The revised passage is easier to read and understand. Short, choppy sentences have been combined, and related points have been pulled together. The same information is conveyed using fewer words and a more organized style.

If the Materials and Methods section is longer than several paragraphs and involves lengthy descriptions of several topics, you may wish to use *subheadings* (perhaps taken directly from your outline) that break the text into clearly labeled sections (for example, Study Area, Sampling Methods, and Data Analysis; or Test Water and Fish, Testing Conditions, Chemical Analyses, and Statistical Analyses; or Plant Material, Morphometry, Light Microscopy, and Electron Microscopy). Make your subheadings general or more specific, depending on the type and amount of information you need to relate. If you use many specialized or invented terms to report your results, they may be put in the Methods section under Definitions. Using subheadings makes your text easier to write and to read, and it prods your memory for stray details on all aspects of the study.

■ Use specific, informative language.

Give your readers as much information as you can. Replace vague, imprecise words with more specific ones, and quantify your statements wherever possible.

VAGUE	I observed some monkeys in a large outdoor enclosure and others in small, individual indoor cages.
SPECIFIC	I observed 13 monkeys in an outdoor enclosure ($10 \times 8 \times 12$ m^3) and 12 others in individual indoor cages ($1 \times 2 \times 1$ m^3).
VAGUE	Several pits were dug at each forest site, and soil samples were collected from three different depths in each pit.
SPECIFIC	Four randomly located pits were dug at each forest site, and soil samples were collected from three depths at each pit: 0–5 cm, 6–11 cm, and 12–17 cm.
VAGUE	Root nodule tissue was stained with a number of histochemical reagents.
SPECIFIC	Root nodule tissue was stained with toluidine blue, Schiff's reagent, and aceto-orcein.
VAGUE	Every nest was checked frequently for signs of predation.
SPECIFIC	Every nest was checked twice daily (at 0800 and 1600) for signs of predation on eggs or nestlings.

■ Understand the difference between the active and the passive voice.

In the passive voice, the subject of the sentence *receives* the action:

Lizards (SUBJECT) *were collected* from three different sites.

However, in the active voice, the subject *performs* the action:

I (SUBJECT) *collected* lizards from three different sites.

Decisions about whether to use the passive or the active voice arise most often in the Materials and Methods section of research papers. Use of the passive voice is widespread in scientific writing, even though the active voice is generally more direct and concise. However, the passive voice focuses the reader's attention on the objects being studied or manipulated rather than on the researcher, and for this reason the passive voice often seems more appropriate. Many writers find that a carefully constructed mix of active and passive sentences works best. See Chapter 7 (pp. 189–191) for more information on this subject.

■ Omit unnecessary information.

Include only those procedures directly pertaining to the results you plan to present. Do not get carried away in your desire to include all possible details. The reader is not interested in superfluous details or asides, and in published articles such material just wastes space and raises printing costs. The following example includes many unnecessary details:

> Fathead minnows were collected from Lost Lake, near Holmes, North Carolina, and transported back to the third-floor laboratory in large white pails. On the following Tuesday morning, skin for histological examination was taken from the dorsal part of the fish just behind the anterior dorsal fin.

Do we really need to know that the fish rode to the lab from the lake in large white pails? Or that they were taken to that particular laboratory? Or that histological studies were done on a Tuesday?

Notice how much clearer the following example becomes when superfluous details are removed:

FAULTY   After considering a variety of techniques for determining the sugar content of nectar, I decided to use the method developed by Johnson (2004), because it seemed straightforward and easy to follow, especially for someone with a poor mathematical background.

REVISED   I used Johnson's (2004) method to measure the sugar content of nectar.

Do not confuse readers of your Materials and Methods section by referring to material in your Results section.

> To quantify the fright response, I observed 10 groups of fish, each composed of five individuals, and recorded the number of movements per 5-min interval (Fig. 1).

Remember that you are dealing strictly with procedures here. When readers encounter Figure 1 later in the paper, they will know enough to consult the Methods section if they need to.

RESULTS

■ Summarize and illustrate your findings.

The Results section should (1) *summarize* the data, emphasizing important patterns or trends, and (2) *illustrate and support* your generalizations with explanatory details, statistics, examples of representative (or atypical) cases, and tables and/or figures. To convey the results clearly, your writing must be well organized. Present the data in a logical order, if possible in the order in which you described your methods. Use the *past* tense when describing your own findings (see pp. 191–192). If the Results section is long and includes many different topics, consider using subheadings to make the text easier for the reader to grasp.

The following paragraph is from a student paper on reproductive behavior in a species of damselfly. Notice that the author begins with a general statement and then supports it with quantitative data, a graph accompanying the text, and a selected example.

> Observations of 21 marked males showed that the number of matings per day varied among individuals. The number of matings per male ranged from 0 to 10 per day, with a mean of 6.5 ($s_{\bar{x}}$ = 3.2). Males occupying territories with abundant emergent vegetation encountered more females and mated more frequently than males occupying sparsely vegetated areas (Fig. 3). One male, whose territory consisted entirely of open water, obtained no matings during the five days he was observed there.

■ Do not interpret the data or draw major conclusions.

The Results section should be a straightforward *report* of the data. Do not compare your findings with those of other researchers, and do not discuss why your results were or were not consistent with your predictions. Avoid speculating about the causes of particular findings or about their significance. Save such comments for the Discussion.

The following passage is from the Results section of a student paper on the algae present in a small stream. The writer begins with a statement of her findings (the first sentence), but then continues with interpretive material that really belongs in the Discussion.

> The epilithic community was dominated by *Achnanthes minutissima* (Table 4). The abundance of this species at the study site may be related to its known tendency to occur in waters that are alkaline (Patrick 1977) and well oxygenated (DeSeve and Goldstein 1981). Its occurrence may also have been related to...

Although generally biologists save interpretations of the data for the Discussion, you should note that some journals do allow a combined Results and Discussion section, followed by a short Conclusions section. In some studies, particularly those involving a series of successive experiments, readers can more easily understand the author's findings if one set of data is explained and interpreted before presentation of the next set. In such cases, the need for clarity should override a rigid adherence to conventional format. If an academic assignment seems to call for this sort of organization, check with your instructor.

■ Integrate quantitative data with the text.

If the Results section includes tables or figures (discussed fully in Chapter 3), be sure to refer to *each* of these in the text. Do not excessively repeat in the text what is already shown in a table or figure, but also don't restrict yourself to passing comments ("Results are shown in Table 1"). Instead, point out the most important information or patterns and discuss them in the context of related data.

> Reproductive activity was closely related to time of day. Both males and females began to arrive at the breeding site in late morning, and the density of both sexes was highest between 1300–1500 (Fig. 3). Data on air temperatures (Fig. 4) suggest that . . .

Do not automatically assume that anything involving numbers *must* be tabled or graphed in order to look important or "scientific." Unnecessary tables or figures take up space and waste the reader's time. Once you are sure which results are important and why, you may find that many can be summarized easily in a sentence or two. Following are some general guidelines about how to present numerical results verbally.

1. A mean value $(\overline{X})$ is often accompanied by the standard deviation (SD), which gives the reader a sense of the variability of the data within the sample. Where consideration of the highest and lowest values is important, the range may be reported along with the sample size (number of observations).

If you wish to show how reliable a sample mean is as an estimator of the population mean, give the standard error. See Chapter 2 (pp. 39–44) and Additional Readings (pp. 253–255) for useful references on the use of these and other basic statistics.

The following sentences illustrate how statistical results can be integrated smoothly with the text:

> The 15 caterpillars in Group 3 averaged 2.1 cm in total length (range = 1.0).

> In the bullfrog choruses at Werner Lake, the mean size of central males was 140.42 mm (SD = 7.45, n = 12).

You may also denote the standard deviation as follows:

140.42 ± 7.45 mm

2. When reporting the results of commonly used statistical analyses, you need not describe the tests in detail nor give all your calculations.

Generally, you need to report only the major test statistics, along with the significance or probability level (see pp. 87–88).

> Analysis of variance showed significant variation among females with respect to mean egg mass ($F_{[29.174]}$ = 25.4, $P < 0.001$).

3. As shown above, conventional abbreviations and symbols are used to report data succinctly within sentences. Some common ones are listed on the inside back cover of this book.

4. When do you write out the words for numbers, and when do you use numerals? The Council of Science Editors (CSE) recommends the use of arabic numerals for all measurable and countable entities. For example, you should use numerals when you (1) report statistics; (2) give quantitative data using standard units of measurement; and (3) refer to dates, times, pages, figures, and tables:

> The experiments were conducted between 1200 and 1500 on September 8, 9, and 10, 2005.

> Overlap of burrow groups ranged from 0 to 35% and averaged 12% (Fig. 2 and Table 4).

> We marked 14 solitary male dragonflies, 9 females, and 15 pairs.

Do not *begin* a sentence with a numeral. Either write out the number or revise the sentence. If the number is part of a chemical term, it cannot be spelled out and the sentence should be reworded.

Faulty	12 out of 425 eggs were cracked at the start of the first experiment.
Revised	Twelve out of 425 eggs ...
Faulty	6-mercaptopurine was used to inhibit mitosis.
Reworded	Mitosis was inhibited by 6-mercaptopurine.
Reworded	I used 6-mercaptopurine to inhibit mitosis.

When one numeral is adjacent to a second numeral, spell one of them out for clarity, preferably the numeral that is not linked with a unit of measurement. Alternatively, reword the sentence.

Faulty	I conducted 10 15-min observation sessions ...
Revised	I conducted ten 15-min observation sessions ...
Reworded	I conducted 10 observation sessions, each lasting 15 min ...

The CSE recognizes that it may be preferable to use words rather than numerals when the primary intent or context of the passage is *not* quantitative. This situation sometimes arises with the use of "one" (as opposed to "1"):

These data suggest that *one* of the most influential factors . . .

According to *one* recent model . . .

In the above examples, the writer's meaning is more qualitative than quantitative. By contrast, it is more appropriate to use the numeral 1 in the following sentence, where exact counts of different entities are reported:

Of the 11 marked dragonflies, 8 were resighted the same day, 2 were never observed again, and 1 died as a result of handling.

When "one" is used as a pronoun, it is always written out:

When comparing these two studies, *one* might argue that . . .

With respect to ordinal numbers, the CSE recommends spelling out single-digit ordinals (*first* through *ninth*) and using numeric form for larger ordinals:

The third example . . .

The 19th and 20th subjects . . .

CSE style advises against the use of superscripts in ordinals: thus, 19th should be typed as 19th, as shown above. Note that some word-processing programs automatically produce superscripts in such situations, so turn off automatic superscripting and proofread your text carefully.

In CSE style, use the numeric form for everything if you have a mix of single-digit and higher ordinals:

In the 5th instance . . . while in the 10th instance . . .

See the CSE manual (2006) for more details about the use of numerals as opposed to words in scientific writing.

5. Watch your wording when you report quantitative results. Several commonly used words have restricted statistical meanings when they appear in scientific writing; do not use them loosely.

Significant: In popular usage, this word means "meaningful" or "important." Scientists use this term in a more restricted sense to refer to *statistically significant* (or nonsignificant) results, after having conducted appropriate statistical tests:

Observed frequencies of turtles obtaining food differed *significantly* from expected frequencies ($X^2 = 58.19$, df $= 8$, $P < 0.001$).

Because the above results were significant, we know that the difference between observed and expected frequencies is probably a valid one. In this case, the likelihood of a difference of at least this magnitude occurring by chance alone is < 0.001, or less than one in a thousand. Note that often the leading zero is implied but not typed in probability values. Thus < 0.001 is often given as $< .001$. However, the CSE prefers the inclusion of an initial zero preceding the decimal point for values smaller than 1.

When reporting the results of statistical tests, you always need to specify the significance level, as indicated by a probability value (P) such as the one above. Results associated with probabilities greater than 0.05 are generally considered *not* significant.

Correlated: In popular usage, two entities that are correlated are related to each other in some way. In scientific writing, the term *correlated* is used in conjunction with certain statistical tests (correlation analyses) that provide a measure of the strength of relationship between two numeric variables.

> Female size was not significantly *correlated* with the percentage of abnormally developing embryos in an egg mass ($r = 0.29$, $P < 0.05, n = 23$).

Note that correlation does *not* allow you to automatically assume a cause-and-effect relationship between the variables. It merely describes the extent to which they covary. For example, if variables X and Y are correlated, above average values of X tend to be associated either with above average values of Y (positive correlation) or with below average values of Y (negative correlation).

Random: In popular usage, this word means "haphazard" or "with no set pattern." However, scientists use this term to refer to a *particular* statistically defined pattern of heterogeneous values. If you write, "Subjects were assigned *randomly* to either Group A or Group B," this implies that you have used a table of random numbers or some other accepted method to make your group assignments truly random.

■ Omit peripheral information and unnecessary details.

Most scientists amass far more data than they ever present to their readers. Similarly, even in a weeklong project, you may have more results than you know what to do with. You may, understandably, feel reluctant to part with any of them. However, if you pack every last bit of information into the paper, you may lose sight of why you did the study in the first place. You must learn to present results selectively, to choose the data that are relevant to your hypothesis because they are either consistent or inconsistent with it. The data you present as results are the same data you will use to support your conclusions. You do not want to confuse the reader (or yourself) by including irrelevant information.

For example, the passage below is from a student field project on the diversity of vascular plants at a small pond.

```
As shown in Figure 1, the shoreline of Hicks
Pond was generally predominated by grasses and
sedges. Cattails occupied a small area on the
northern end, and goldenrods (Solidago spp.)
were present in scattered groups along the east-
ern shore. A large population of red-spotted
newts (Motopthalmus viridescens) was also spot-
ted in this area, along with several species of
ducks. Species diversity was highest at Site D
(see Table 1), which was drier than most areas
and contained some of the same plants as in the
surrounding hayfield.
```

The third sentence in this paragraph should be omitted, because such zoo-logical observations, however interesting, are irrelevant in this context.

Similarly, do not clutter the Results (or any other) section of the paper with irrelevant general statements of aims or intent. The following sentence, for example, should have been omitted from the student paper in which it appeared:

```
To present the results of this study, I will
first examine all of the relevant physiological
factors and then discuss the findings of each
feeding experiment.
```

You need not explain how you will proceed. If your writing is coherent and well organized, readers will follow your train of thought.

DISCUSSION

■ Interpret your results, supporting your conclusions with evidence.

In the Results section you reported your findings; now, in the Discussion, you need to tell the reader what you think your findings mean. Do the data support your original hypothesis? Why or why not? Refer to your data (in the *past tense*), citing tables or figures where necessary; use these materials as evidence to support your major argument or thesis. Here is the place, too, to discuss the work of other researchers. Are your findings consistent with theirs? How do your results fit into the bigger picture? Use the *present tense* when discussing the published work of other researchers (see pp. 191–192).

For many students, the Discussion section of a research paper is the most difficult portion to write. If you find yourself floundering, you may not fully understand what you are writing about yet. Take some time to rethink your study and why you did it. Look again at relevant literature on your topic. Then try drafting your ideas again. One of the chief benefits of writing and rewriting is to articulate your ideas ("writing to learn"). Be sure to leave ample time for this important process.

■ Do not present every conceivable explanation.

Sometimes beginners feel obliged to think of every possible way to interpret their results. The reader, swamped by explanations (most of them highly speculative), will quickly lose faith in the author. Look at this excerpt from a student Discussion section:

> The dramatic decrease in *Ochromonas* in December
> may have been related to the formation of cysts
> (perhaps overlooked in the water samples taken
> at that time). It is also possible that this
> alga died off suddenly because of some environ-
> mental stress, such as a prolonged period of
> cold temperatures or a sudden chemical change in
> the water. Light intensity and photoperiod are
> other potential factors. Finally, the December
> samples were not analyzed until the day after
> they had been collected because two members of
> our lab group were sick, so by this time the
> algal populations may have been different from
> the ones at the field site.

Remember that your task in the Discussion is to argue on behalf of the most plausible interpretations, based on the evidence available to you. To do so, you must be selective, focusing on explanations that have the greatest bearing on your study. Omit wild guesses and irrelevant asides.

■ Recognize the importance of "negative" results.

Experiments do not always have to confirm the presence of major differences, a strong relationship between two variables, or a conspicuous trend or pattern to be considered good science. Sometimes you may find no significant difference between two groups or no relationship between a particular process and, say, temperature or some other factor you have been investigating. Such "negative" results still constitute respectable science, and they still need explanation. Thus, be alert to unexpected findings, and don't automatically conclude that the experiment is a failure or that you've made a mistake. Cases that do not conform to the expected pattern might represent something important — perhaps a new or altered focus for your study.

■ Make your prose convey confidence and authority.

Show that you are knowledgeable about your subject and take responsibility for your conclusions. Do not hedge or apologize. The tone in the following student examples is unnecessarily tentative:

```
This is just a preliminary study of the social
behavior of the guppy (Poecilia reticulata), but
it may possibly shed some light on the subject
and serve as a base for more conclusive work in
the future.
```

Would you want to read this paper? Perhaps not, if it is as preliminary and inconclusive as the author implies.

```
These results seem to suggest the possibility of
at least some connection between ferric chloride
and increased disease resistance.
```

This writer seems afraid to take a stand. What *do* the data show? If they suggest a connection between ferric chloride and disease resistance, then the writer should say so: "These results suggest a connection between ferric chloride and disease resistance."

```
These findings may not be very accurate because
of my limited experience, although they appear
to be consistent with the observations of Henry
(1999) and Blanksteen (2003).
```

The last writer, like many other students, obviously feels intimidated by the work of "experts." However, "beginners" often do very good scientific work. Do not weaken your conclusions by unnecessary references to your status as a novice. If you are reluctant to make a strong statement, ask yourself why. Perhaps your anxiety stems from reservations about the data or the way the experiment was conducted. Perhaps there are certain variables you have not considered or certain assumptions that don't ring true.

Those factors *are* worth considering, and it is important to sort them out from any general insecurities you have about yourself as a scientist or a writer.

■ Use a coherent, logical organization.

Instead of proceeding from the *general* to the *specific,* as the Introduction does, the Discussion moves from the *specific* to the *general.* There is no one right way to put together a Discussion, but the following plan is a common and effective one.

1. Start by drawing attention to your major findings without excessively repeating the Results. The reader has just looked at your data, so you do not have to describe them all over again. Beginning writers often start the Discussion by dredging up material from the Introduction; this approach, too, just adds redundancy to the text.

Focus the reader's attention on the most important findings, patterns, or trends. For example, here is the beginning of the Discussion section from a paper by Young and others (1986, p. 400):

> Although interactions between molluscs and their potential predators have been studied extensively (for reviews see Thompson 1960, Edmunds 1966, Todd 1981, Faulkner and Ghiselin 1983), these data are the first in which predator-prey interactions have been documented for a large number of potential predators of intertidal Onchidiids. In addition, our data indicate that the defensive secretion of *Onchidella borealis* has an effect on the distribution of the predatory seastar *Leptasterias hexactis.* The results indicate that *O. borealis* does not fire its repugnatorial glands in response to all potential predators, nor do all potential predators demonstrate flight behaviors in response to the glandular secretions of *O. borealis.*

2. Ask yourself what causes may underlie the major trends or phenomena you have described in the paper. If there are conflicting, problematic, or unexpected results, suggest explanations. Here is an excerpt from a student research report:

> The unusual presence of *Chaenorrhinum minus* in an untended garden (Site 5), instead of in its usual habitat of railroad cinder ballast, may be related to the use of railroad ties as decorative borders around the garden. Seeds from *C. minus* may have been lodged in crevices in the railroad ties, transported to the garden, and dislodged as the ties were being set in place.

However, it is still unusual that *C. minus* was
growing so well in a moist, shaded area among
many other plants. My findings suggest that
C. minus generally is a relatively poor competi-
tor and grows best on dry, exposed, gravelly
sites where few other plants are found.

3. Compare your findings with the work of other researchers. Are
your results similar to theirs? At this point you can begin to supplement
your own evidence with relevant findings from other studies, showing the
reader how your work is part of a broader framework. Be sure to look at
the subject fairly and honestly. If some authors obtained results different
from yours, point this out and suggest explanations for the differences. The
following example is from McMillan and Smith (1974, p. 52).

The activities of males after spawning interested other observers of
fathead minnows. Miller (1962) reported that a parental male positioned himself in the mouth of the cavity below the eggs and constantly fanned with his pectoral fins. This behavior has not been
noted by other workers, nor has it been observed during the present study. It is possible that Miller was watching males hovering
(our definition) below their eggs. In that case it is doubtful that
the weak pectoral fin movements employed in hovering could
substantially aerate the eggs—the function Miller attributed to
"fanning."

4. End with more far-reaching predictions, interpretations, and conclusions. Can you generalize from your specific findings to other situations? How does your work contribute to an understanding of the broader
topic? If you can end with a firm statement, as the student example below
does, you give the reader a satisfying sense of closure.

In conclusion, cannibalism of eggs by larvae of
the butterfly *Euphydryas phaeton* occurred commonly under natural conditions, even within relatively small colonies. These results are in
agreement with Wilson's (1975) suggestion that
detailed studies of cannibalism in animals may
show it to be more common than is usually sup-
posed. The finding that cannibalistic larvae
grew more rapidly than noncannibalistic ones
suggests that cannibalism may be an important
factor in larval development, especially when
food supplies are scarce. This nutritional

```
benefit of cannibalism may have long-reaching
effects. If successful laboratory culture meth-
ods can be developed for E. phaeton, we can
further explore the relationship between canni-
balism and individual fitness.
```

ACKNOWLEDGMENTS

A short Acknowledgments section usually comes between the Discussion and the Literature Cited or References sections in a published research paper. In this section you list sources of funding for the project and thank anyone who assisted you with your research or with the preparation of your paper. For more on writing an Acknowledgments section, see pages 204–205 in Chapter 8.

LITERATURE CITED

In the Literature Cited or References section, you list all the references *cited* (referred to) in your paper. Do not list any other sources, even if they proved to be useful background reading. See Chapter 6 for format guidelines.

CHECKLIST FOR LAB REPORTS AND RESEARCH PAPERS

Title

- Is it specific and informative?
- Does it contain relevant keywords?
- Does your title accurately reflect the contents of your paper?

Abstract

- Are you concise?
- Do you summarize all major parts of the paper?
- Have you omitted literature citations?
- Can your Abstract stand alone and still make sense?

Introduction

- Do you give a brief overview of the topic, providing a context for your study?
- Have you cited pertinent references to the literature, using an appropriate scientific documentation style?
- Do you state your objectives, hypothesis, or predictions?

Materials and Methods

- Have you explained all the procedures used for collecting the data presented in your Results?
- Is there enough information to enable someone else to repeat your study?
- Have you omitted any unnecessary details?

Results

- Do you summarize all major findings, supporting your points with specific references to your data?
- Are your results supported by appropriate statistical analysis?
- Do you provide a coherent narrative, using the past tense?
- Have you omitted peripheral or unnecessary details?
- Do you refrain from interpreting your findings and from citing other literature?
- Is every table and figure absolutely necessary?
- Does every table have a title, every figure a legend?
- Do you use a figure rather than a table to show a pattern, relationship, or trend?
- Are data in related figures or tables presented in a consistent manner?
- Are data in figures, tables, and text consistent with one another?
- Are figures and tables numbered consecutively in separate series?
- Do you refer to every figure and table, and have you numbered them in the order in which they are mentioned in the text?
- Are tables and figures understandable apart from the text?

Discussion

- Do you address the major implications of your findings?
- Have you considered problems, inconsistent results, and counterevidence?
- Have you interpreted your results in the context of other literature?

- Do you use the past tense to refer to your own findings and the present tense to refer to established literature?
- Is your scientific documentation style the same as the one used in the Introduction?
- Have you cited all important sources?
- Is your discussion coherent and organized?

Acknowledgments

- Have you thanked all those who helped with the research or writing of the paper?
- Have you acknowledged any sources of funding for the project?

References / Literature Cited

- Are all sources cited in the text also listed in your References?
- Does your References section contain any sources not cited in the text?
- Have you followed a conventional format consistently and meticulously?
- Is the information about every source accurate and complete?

SAMPLE LABORATORY REPORT

Following is a sample lab report for an introductory botany course. This report addressed a specific course-based exercise designed to introduce students to plant sampling methods. The instructor presented pairs of students with the same objective: to determine the distribution pattern of dandelions on a section of the college campus. He provided the class with several references to consult in writing their lab reports and also specified a particular scientific documentation format, a variation of the name-year method. Here, the intended audience includes fellow classmates and, particularly, the course instructor.

Distribution Pattern of Dandelion
(Taraxacum officinale)
on an Abandoned Golf Course

Marin Johnson
Laura Arnold

Lab 4
Botany 100A
September 13, 200-

Include title,
authors, course,
and date on
title page for
lab report.
Do not number
title page, but
consider it
page 1. Use
page numbers
for all subse-
quent pages.

INTRODUCTION

Abstract not required for this assignment. Introduce topic and refer briefly to pertinent literature. Name-year method is used here to cite sources. Use present tense to refer to established knowledge.

State specific objectives of project.

Theoretically, plants of a particular species may be aggregated (clumped), random, or uniformly distributed in space (Ketchum 2005). The distribution type may be determined by many factors, such as availability of nutrients, competition, distance of seed dispersal, and mode of reproduction (Kershaw and Looney 1985).

The purpose of this study was to determine if the distribution pattern of the common dandelion (*Taraxacum officinale*) on an abandoned golf course was aggregated, random, or uniform.

METHODS

The study site was an abandoned golf course on the campus of Ford College, Hilton, New York. The vegetation was predominantly grasses, along with dandelions, broad-leaf plantain (*Plantago major*), and bird's-eye speedwell (*Veronica chamaedrys*). We sampled an area of approximately 6 ha on 10 July 2005, approximately two weeks after the golf course had been mowed.

To ensure random sampling, we threw a tennis ball high in the air over the study area. At the spot where the tennis ball came to rest, we placed one corner of a 1 × 1 m^2 metal frame (quadrat). We then counted the number of dandelion plants within this quadrat. This procedure was repeated for a total of 111 randomly-placed quadrats.

Describe procedures in enough detail for others to repeat study. Include statistical methods used.

We used the two-step procedure described by Kershaw and Looney (1985). We first tested whether the distribution of dandelion was random or non-random. From the counts of the number of dandelions in our 111 quadrats, we used a log-likelihood ratio (G) test to examine the goodness of fit between our observed frequencies and those expected based on the Poisson series $e^{-\mu}$, $\mu e^{-\mu}$, $\mu^2/2!e^{-\mu}$, $\mu^3/3!e^{-\mu}$, . . . , where μ is the mean density of plants per quadrat. In carrying out this test, we grouped observed and expected frequencies so that no group had an expected frequency less than 1.0 (Zar 2005). We then determined whether the distribution was aggregated or uniform by calculating the coefficient of dispersion (ratio of the variance to the mean). A coefficient $>$ 1 indicates an aggregated distribution whereas a coefficient $<$ 1 indicates a more uniform distribution. Finally, we tested the significance of any departure of the ratio from a value of 1 by means of a t-test.

RESULTS

Table 1 shows the number of quadrats containing 0, 1, 2, . . . , 17 dandelion plants. More than two-thirds (67.6%) of the 111 quadrats contained no dandelion plants; almost 90% (89.2%) of the quadrats contained fewer than 3 dandelion plants. We observed a highly significant lack of fit between our observed

Report your major findings, using past tense. Refer to table, but do not simply repeat its contents. Use "significant" only for statistical significance.

Depending on instructor's preference, insert table into text close to where it is first mentioned, or else put at end of paper.

Table 1. Frequency distribution of dande-lion (*Taraxacum officinale*) plants in 1×1 m^2 quadrats positioned randomly over a 6-ha area on an abandoned golf course. Expected frequencies were calcu-lated from the successive terms of the Poisson distribution (see Methods).

no. per quadrat	observed frequency (f_i)	expected frequency ($\hat{f}_i$)
0	75	38.68594
1	12	40.77707
2	12	21.49062
3	2	7.550757
4	3	1.989727
5	2	0.419456
6	0	0.073688
7	2	0.011096
8	0	0.001462
9	1	0.000171
10	0	1.8×10^{-5}
11	0	1.73×10^{-6}
12	0	1.52×10^{-7}
13	1	1.23×10^{-8}
14	0	9.27×10^{-10}
15	0	6.52×10^{-11}
16	0	4.29×10^{-12}
17	1	2.66×10^{-13}
Total	111	

frequencies and expected frequencies based on the Poisson distribution ($G = 78.4$, df = 3, $P < 0.001$). Thus, our data indicated that the distribution pattern of dandelion plants on the abandoned golf course was not random. The mean number of dandelion plants per quadrat was 1.05 (SD = 2.50), and the coefficient of dispersion was 5.95. A t-test showed that this value is significantly greater than 1.0 ($t = 36.7$, df = 110, $P < 0.001$), which strongly supports an aggregated distribution of the dandelion plants.

DISCUSSION

An aggregated (clumped) distribution is the most commonly observed distribution type in natural populations (Begon and others 1996). Among plants, aggregated distributions often arise in species that have poorly dispersed seeds or vegetative reproduction (Kershaw and Looney 1985). In the dandelion, the seeds are contained in light, parachute-bearing fruits that are widely dispersed by the wind. This method of seed dispersal would tend to produce a random distribution. However, dandelion plants also reproduce vegetatively by producing new shoots from

Interpret results and link them to pertinent literature. Acknowledgments section not necessary for this assignment.

existing taproots, and what we considered as groups of closely-spaced separate individuals probably represented shoots originating from the same plant. Thus, vegetative reproduction probably accounted for the observed aggregated distribution in this species.

REFERENCES

Begon M, Harper JL, Townsend CR. 1996. Ecology: individuals, populations and communities. Oxford: Blackwell Science Limited. 1068 p.

Kershaw KA, Looney JHH. 1985. Quantitative and dynamic plant ecology. 3rd ed. London: Edward Arnold. 282 p.

Ketchum J. 2005. Lab manual for Botany 100. Ford College, Hilton, NY. 63 p.

Zar JH. 2005. Biostatistical analysis. 5th ed. Englewood Cliffs (NJ): Prentice-Hall. 960 p.

Include all references cited. Alphabetize sources by first author's last name. Follow guidelines given by instructor.

Cite course materials, if appropriate.

SAMPLE STUDENT RESEARCH PAPER

The following paper describes a more advanced student research project in plant physiology. In this case, the class worked together for several weeks to collect and analyze the data. Each student was then responsible for presenting these results in an independent research paper appropriate for a scientific journal. The student was required to consult at least eight primary sources and to follow CSE name-year documentation style. Note that, despite these differences in course requirements and instructor expectations, both sample research reports follow the typical format of a scientific paper.

Germination of *Arabidopsis thaliana* in
Response to Gibberellins in
Light and Darkness

Kristin A. VanderPloeg

Biology 251
April 21, 200-

Include title,
author, course,
and date on
title page for
student paper.
Do not number
title page, but
consider it
page 1. When
submitting a
manuscript for
publication, fol-
low journal
format require-
ments.

Germination of *Arabidopsis thaliana* 2

ABSTRACT

Seed germination in plants is controlled by several mechanisms, including the presence of gibberellins as well as environmental conditions, such as temperature, light, and moisture. Germination of dormant seeds can be promoted by application of gibberellins. These hormones apparently act as natural endogenous regulators of dormancy, but their precise role is unclear. Here I report the effect of exogenous GA_3 on the germination of seeds of the *gal-3* mutant of *Arabidopsis thaliana* in both light and darkness. Germination success increased significantly from 20.0% to 97.8% as the concentration of exogenous GA_3 increased from 0.001 to 0.5 mol L^{-1} in both light and dark conditions. However, there was no significant difference in germination success between the light and dark treatments. These data support the hypothesis that gibberellins are critical regulators of seed germination but are not consistent with previous studies showing more germination in light than in darkness.

Keywords: gibberellin, gibberellic acid, germination, *Arabidopsis thaliana*, seed, dormancy, light sensitivity.

INTRODUCTION

Buried seeds of numerous plant species undergo a period of dormancy in response to environmental influences such as temperature and lack of sufficient

Italicize genus and specific epithet. Capitalize genus only.

Summarize paper's most important contents using past tense. Use present to suggest a general conclusion.

Label and begin on same page as Abstract. In longer paper or journal manuscript, start new page.

Germination of *Arabidopsis thaliana* 3

moisture. In *Arabidopsis thaliana* (L.),
dormancy allows seeds to survive the dry
summer and germinate in autumn (Leon-
Kloosterziel and others 1996). In an
agricultural context, the establishment
and subsequent breaking of dormancy may
affect the efficiency of crop yield and
the preservation of seeds for later
planting. As reviewed by Baskin and
Baskin (1998), factors regulating dormancy include light, temperature, and
hormones such as abscisic acid and gibberellins.

   Gibberellins (GAs) exert profound
effects on plant development. Numerous
studies support the involvement of GAs
in releasing seeds from dormancy (see,
for example, Karssen and others 1989;
Arnold and others 1996). Various forms
of GAs have been identified, and it is
evident that the bio-active GA depends on
the species of plant being examined
(Derkx and others 1994). In *A. thaliana*,
a pathway converts inactive GA_9 and GA_{20}
to bio-active GA_4 and GA_1, respectively
(Yamaguchi and others 1998). The nature
of this conversion is the result of the
expression of GA 3ß-hydroxylase (Yamaguchi
and others 1998).

   GA_1 and GA_4 may facilitate germination
by furthering the penetration of the seed
coat by the radicle (Yamaguchi and others
1998) or by producing an enzyme responsible for digesting tissue surrounding the

Spell out *Arabidopsis* once,
then abbreviate
it when preced-
ing species
name, except at
beginning of a
sentence.

For back-
ground, briefly
review perti-
nent literature.
CSE name-year
method is used
here to cite
sources. Use
present tense to
refer to estab-
lished knowl-
edge.

Germination of *Arabidopsis thaliana* 4

radicle tip (Groot and Karssen 1987). Recently, a gene, (*RGL2*) that regulates seed
germination in response to GAs has been
identified in the *Arabidopsis* genome (Lee
and others 2002). In addition, SPINDLY, a
negative regulator of GA responses, has
been found in *Arabidopsis* (Greenboim-Wainberg and others 2005). This factor also
represses plant responses that depend on
cytokinins, indicating that some interaction between these hormones and GA responses is likely.

Further understanding of the role of
gibberellins in seed germination has been
made possible by the use of gibberellin-
deficient (*ga1*) mutants of *A. thaliana*.
The mutation in *ga1* strains prevents synthesis of active GAs (Zeevart and Talon
1992); the wild-type phenotype is re-
stored by application of exogenous GAs.

The involvement of light in the
biosynthesis of GAs has also been studied. GAs can mimic the effects caused by
light in seed development and promote
germination, and exposure to light promotes the expression of genes responsible
for the synthesis of GA 3ß-hydroxylase
(Yamaguchi and others 1998). In the absence of this enzyme, gibberellins GA_9 and
GA_{20} are present but are not converted to
physiologically active forms. Indirectly,
therefore, light stimulates the synthesis
of bio-active GAs and thus promotes germination of seeds of *A. thaliana*. This

Germination of *Arabidopsis thaliana* 5

finding is supported by work by Derkx and
Karssen (1993a), who found that, in the
presence of light, the minimum concentration of a mixture of GA_4 and GA_7 needed to
produce germination in wild-type and *gal-2*
mutants of *A. thaliana* is lower than the
amount needed in darkness. Thus, light
increases the sensitivity of seeds of
both varieties of *A. thaliana* to GAs.

I examined the effect of exogenous
GA_3 in light and in darkness on seeds of a
different mutant of *A. thaliana, gal-3*,
which is also dependent on exogenous GAs
for germination.

> After giving rationale for study, state paper's objectives.

MATERIALS AND METHODS

I obtained seeds of *Arabidopsis
thaliana*, mutant *gal-3*, from the Arabidopsis Biological Resource Center,
Columbus, Ohio. Solutions of gibberellic
acid (GA_3) were prepared at concentrations
of 0.001, 0.01, 0.1, and 0.5 mol L^{-1}.
Working under fluorescent illumination, I
added 1 mL of each GA_3 solution or water
to small petri dishes, each containing a
layer of filter paper and 10 *A. thaliana*
seeds. I prepared 19 sets of these 5
treatments. Of these, 10 sets were kept
under fluorescent "Gro-Lights," and 9
sets were kept in a light-tight box. The
number of germinated seeds was counted
every 24 hours for 7 days. Radicle protrusion was taken as the criterion for
successful germination. Water was added
as needed during the 7-day incubation

> Use active voice when appropriate; use passive voice to focus attention on materials, not yourself.

> Describe procedures in enough detail for others to repeat study.

Germination of *Arabidopsis thaliana* 6

period to keep all seeds as close to the same moisture level as possible.

RESULTS

In the light, germination success of seeds of the *ga1-3* mutant increased from 6% to 97.8% with increases in GA_3 concentration from 0 to 0.5 mol L^{-1} (Fig. 1). A similar trend with increasing GA_3 concentration was observed in seeds maintained in darkness.

Report statistical and other data. Use "significant" only for statistical significance.

A two-way analysis of variance (Table 1) showed that the effect of GA_3 concentration on germination success was highly significant ($P < 0.001$). However, there was no significant difference ($P > 0.05$) in germination success between seeds kept in the light and those kept in darkness. In addition, there was no significant interaction between GA_3 treatment and light conditions; thus, the significant effect of GA_3 treatment was the same in both light and darkness. Pairwise comparisons by means of Tukey tests showed that each successive increase in GA_3 concentration throughout the entire range produced a significantly greater germination success in both light- and dark-treated seeds.

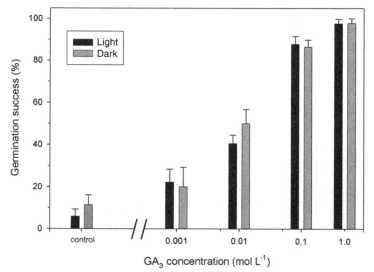

Germination of *Arabidopsis thaliana* 7

Depending on instructor's preference, insert figures or tables into text, close to where they are first mentioned, or else group them at end of paper.

Figure 1. Germination success of seeds of the *gal-3* mutant of *Arabidopsis thaliana* in response to treatment with GA_3 in light or darkness.

Table 1. Two-way ANOVA on factors affecting germination success of seeds of the *gal-3* mutant of *Arabidopsis thaliana*

Source of variation	df	MS	F
GA_3 concentration	4	19340.068	105.96[***]
Light condition	1	74.459	0.408[ns]
GA_3 concentration × light condition	4	87.301	0.478[ns]
Error	85	182.520	

[***] = $P < 0.001$: *ns* = nonsignificant

Germination of *Arabidopsis thaliana* 8

DISCUSSION

Germination of seeds of the *gal-3* mutant of *A. thaliana* increased with increasing concentrations of exogenous gibberellin. Similar findings have been reported for wild-type and *gal-2* mutants of *A. thaliana* (Derkx and Karssen 1993a, 1993b; Derkx and others 1994) and with dormant seeds of *Chaenorrhinum minus* (Arnold and others 1996). All of these authors have concluded that, under natural conditions, endogenous GAs may be responsible for breaking seed dormancy and stimulating germination in wild-type seeds.

Briefly review pertinent findings. Do not excessively repeat results.

My results showed that seeds germinated equally well in light and in darkness at all GA concentrations tested and in the control. However, similar studies (Derkx and Karssen 1993a; Derkx and others 1994) show that light increases endogenous GA levels and, in turn, increases germination success of seeds of the *gal-2* mutant of *A. thaliana*. Further work is necessary to determine reasons for the discrepancy between my results and those cited above for the *gal-2* mutant. One possible explanation is that seeds of the *gal-2* and *gal-3* mutants differ in light sensitivity. It is also possible that germination in darkness was a response to the 20-30 min exposure to fluorescent light during handling. Derkx and Karssen (1993b) have found that an

Relate your findings to those of others. Shift to present tense to discuss previously published information.

Germination of *Arabidopsis thaliana* 9

exposure to red light for as little as 15 min, followed by incubation in darkness, can promote increased germination success. Peng and Harberd (1997) suggest that this sensitivity is due to the presence of phytochromes, which are light-sensitive plant pigment molecules that regulate many light-dependent processes in plants, including seed dormancy and germination. A similar phytochrome-mediated response may explain the findings I report here.

ACKNOWLEDGMENTS

I thank C. LaFave for advice during the research and preparation of this paper.

LITERATURE CITED

Arnold RM, Slyker JA, Gupta TH. 1996. Germination of *Chaenorrhinum minus* seeds in response to gibberellin treatments. J Plant Physiol. 148(6):677-683.

Baskin CC, Baskin JM. 1998. Seeds: ecology, biogeography, and evolution of dormancy and germination. New York: Academic Press. 666 p.

Derkx MPM, Karssen CM. 1993a. Effects of light and temperature on seed dormancy and gibberellin-stimulated germination in *Arabidopsis thaliana:* studies with gibberellin-deficient and -insensitive mutants. Physiol Plant. 89(2):360-368.

With their permission, acknowledge those who assisted with the project.

Include all references cited. Alphabetize sources by first author's last name. Here, follow CSE format.

For a book, give author(s), publication date, title, place of publication, publisher, and number of pages.

For a journal article, use initials for authors' first and middle names followed by publication date, title of paper with only first word capitalized. Use conventional abbreviations for journal title. Give volume and issue number, followed by page numbers.

Germination of *Arabidopsis thaliana* 10

Derkx MPM, Karssen CM. 1993b. Variability
    in light-, gibberellin- and nitrate
    requirement of *Arabidopsis thaliana*
    seeds due to harvest time and condi-
    tions of dry storage. J Plant Phys-
    iol. 141(5):574-582.

Derkx MPM, Vermeer E, Karssen CM. 1994.
    Gibberellins in seeds of *Arabidopsis
    thaliana:* biological activities,
    identification and effects of light
    and chilling on endogenous levels.
    Plant Growth Regul. 15(3):223-234.

Greenboim-Wainberg Y, Maymon I, Borochov R,
    Alvarez J, Olszewski N, Ori N, Eshed
    Y, Weiss D. 2005. Cross talk between
    gibberellin and cytokinin: the
    Arabidopsis GA response inhibitor
    SPINDLY plays a positive role in
    cytokinin signaling. Plant Cell.
    17(1):92-102.

Groot SPC, Karssen CM. 1987. Gibberellins
    regulate seed germination in tomato
    by endosperm weakening: a study with
    gibberellin-deficient mutants.
    Planta. 171(4):525-531.

Karssen CM, Zagórski S, Kepczynski J,
    Groot SPC. 1989. Key role for en-
    dogenous gibberellins in the control
    of seed germination. Ann Bot.
    63(1):71-80.

Lee S, Cheng H, King KE, Wang W, He Y,
    Hussain A, Lo J, Harberd NP, Peng J.
    2002. Gibberellin regulates *Ara-
    bidopsis* seed germination via *RGL2*,

Germination of *Arabidopsis thaliana* 11

a *GAI/RGA*-like gene whose expression is up-regulated following imbibition. Genes Dev. 16(5):646-658.

Leon-Kloosterziel KM, van de Bunt GA, Zeevaart JAD, Koornneef M. 1996. *Arabidopsis* mutants with a reduced seed dormancy. Plant Physiol. 110(1):233-240.

Peng J, Harberd P. 1997. Gibberellin deficiency and response mutations suppress stem elongation phenotype of phytochrome-deficient mutants of *Arabidopsis*. Plant Physiol. 113(4):1051-1058.

Yamaguchi S, Smith MW, Brown RGS, Kamiya Y, Sun T. 1998. Phytochrome regulation and differential expression of gibberellin 3ß-hydroxylase genes in germinating *Arabidopsis* seeds. Plant Cell. 10(12):2115-2126.

Zeevart JAD, Talon M. 1992. Gibberellin mutants in *Arabidopsis thaliana*. In: Karssen CM, Van Loon LC, Vreugdenhil D, editors. Progress in plant growth regulation. Dordrecht: Kluwer Academic Publishers. p. 34-42.

CHAPTER 5

Writing a Review Paper

A scientific review paper is a critical synthesis of the research on a particular topic. Biologists read review papers to stay abreast of the current knowledge in a field and to learn about subjects that are unfamiliar to them.

In writing review papers, authors vary in their purposes and scope. For example, we might classify biological review papers into at least five different types.

A *state-of-the-art review* presents an up-to-date, interpretive synthesis of our knowledge of a certain subject or issue, with emphasis on the most recent literature. For example, an author might look at what is currently known about the advantages and disadvantages of a particular surgical method for mastectomy, or a particular chemotherapy regime in the treatment of breast cancer.

A *historical review* examines the historical development of knowledge about a particular topic. It may focus on the emergence of the subject area as an important field of study, the major pioneering work by early contributors, the replacement of outdated theories or perspectives, and the development of improved methodologies. For example, an author might survey our growing understanding of territoriality in birds, beginning with the earliest ideas and (perhaps) misconceptions, and ending with the most recent models and problems.

A *comparison-of-perspectives review* critically examines two or more ways of looking at a particular scientific issue. For example, an author might contrast studies supporting the idea that the transport of sugars in phloem tissue is an active metabolic process with studies supporting the hypothesis that it is a passive process by mass flow.

A *synthesis-of-two-fields review* attempts to provide new insights about one field of study by bringing in literature from another area or discipline.

For example, a reviewer might examine how the language and methodology of a particular branch of microeconomics has shed light on our knowledge of reproductive strategies in insects.

A *theoretical* or *model-building review* surveys the literature on a particular scientific problem—for example, carbon allocation in trees, or cooperative breeding in birds—with the aim of presenting a new theoretical context or an alternative model.

These arbitrary categories are not rigid, of course, and many reviewers ambitiously and successfully fuse more than one approach in their investigation and subsequent expression of a particular topic. Because of possible differences in emphasis, the format of a review is not as standardized as that of a research paper. Most reviews have an Introduction and a Discussion or Conclusions section, and many also have an Abstract. Long reviews may be preceded by a Table of Contents. All end with a lengthy Literature Cited section.

When you are assigned to write a review paper for a biology course, you need to rely on many of the same strategies used by authors of published reviews. Your instructor will probably expect you to address readers similar to yourself and your classmates — people with a background in the broad subject area but without specialized knowledge of your particular topic. Your aim is not only to inform but also to evaluate and interpret. A good review bears the stamp of the writer's own thought processes.

CHOOSING A TOPIC

Do not underestimate the importance of choosing a suitable topic. Ideally, it should be (1) interesting to you; (2) not so broad that it is unmanageable; (3) not so narrow you can't find enough information on it; and (4) not so difficult that you can't fully understand it.

A common approach is to start with a broad topic and do some general reading about it, gradually narrowing it down to a workable size. Suppose you decide to write a paper about orchids. Because orchids are a large, diverse group and much has been written about them, you will need to restrict yourself to some specific aspect of these plants. For example, you might decide to write about the unusual interactions between orchids and their insect pollinators. Eventually you might narrow your topic even further, perhaps confining yourself to only a few species or to several, similar kinds of pollination interactions.

As you narrow your topic and become familiar with the literature, you need to develop a sense of your main objectives. What question will your paper address and from what perspective? Are you shifting toward a particular viewpoint or conclusion that can serve as a main point, or thesis, for the paper? If you limit your scope and define your goals early in the project, your reading and note taking will be more directed and your time will be more productive.

Occasionally the first topic you choose may be *too* narrow and you will have to enlarge it or shift to a different subject altogether. Allow time for false starts, delays, and topic changes. Recognizing a good subject to write about requires thought and effort.

DECIDING ON A TITLE

Like a research paper, a review paper should have a specific and informative title. If your title is vague or too general, it will suggest that you lack a firm sense of your paper's aims and scope. If you are preparing a paper for publication, remember that many indexing and abstracting services rely on keywords in titles in their organization of the journal literature. Your title should therefore also be concise; cut any words that do not serve an important function in describing the paper's contents. Consider the following sets of examples:

VAGUE	Studies on Seed Coats
IMPROVED	The Role of Seed Coats in Seed Viability
VAGUE	The Bromeliaceae: Research Questions and Controversies
IMPROVED	Physiological Ecology of the Bromeliaceae
WORDY	A Review of the Literature on the Functions of the Avian Spleen
IMPROVED	Functions of the Avian Spleen: A Review
WORDY	Strategies of Seed Dispersal by Plants Inhabiting Desert Environments
IMPROVED	Seed Dispersal Strategies in Desert Plants

See also pages 69–72 for more on titles.

WORKING WITH THE LITERATURE

A review paper involves a thorough search of the literature on your topic, followed by careful reading and selective note taking. These parts of the research process are covered in Chapter 1. However, it bears repeating here that constructing a good review requires the ability to be discriminating in your use of materials. Remember that a review paper is a secondary source that provides a critical, focused synthesis of *primary* sources. In biology, the primary literature consists largely of research papers in scientific journals, along with government or technical reports, conference proceedings, and other works presenting original research to a scientific audience. It is perfectly acceptable to *start* your research with scientific encyclopedias, textbooks, lab manuals, nontechnical books, and articles in *Discover,* *National Wildlife, Scientific American, Natural History,* and similar magazines for lay readers. You also may wish to consult reliable online sources for background information. All these sources provide excellent preparation for the more specialized papers you will need to read later. Scientific review papers in your general subject area also serve this function. However, secondary references should not be major sources of information for your paper.

For example, suppose you are writing a review paper on the visual system of horseshoe crabs — a topic you chose after coming across an article titled "What the Brain Tells the Eye" by Robert B. Barlow, Jr., in *Scientific American.* The article contains much information that would be useful for your paper, and like other features in the magazine, it is written by an expert on the topic. However, you need to go one step further. Rather than using the *Scientific American* article as a major source for your paper, you should track down some of Barlow's original research findings in the scientific papers in which they first appeared. For example, in the list of Further Reading following the *Scientific American* article is a paper by Barlow and three coauthors titled "Circadian Rhythms in *Limulus* Photoreceptors: I. Intracellular studies," in *Journal of General Physiology.* With the background knowledge you've acquired by reading Barlow's more general article, you should feel ready to tackle this primary source, along with other specialized research papers. Such references make up the primary literature on vision in horseshoe crabs.

In summary, for a biological review paper you need to focus on papers in scientific journals, along with other primary sources. Rely on secondhand information as little as possible; cite these materials rarely, if at all. If you are unsure about how to distinguish primary from secondary sources or about how to locate primary sources, review Chapter 1 or seek help from your instructor or your reference librarian. With time, you'll become familiar with the important journals featuring research on your topic, and probably, too, with the names of key investigators in the field. Remember that the quality of your review paper will depend largely on the sophistication of your literature search, your ability to make informed choices among a variety of possible references, and your mastery of key primary literature.

PRESENTING YOUR MATERIAL

■ Sketch out a rough plan or make an outline.

Biological review papers are not as standardized in their format as research papers; their organization depends on the subject, the writer's objectives, and (in the case of published articles) the editorial guidelines of the journal. Most reviews have an Introduction, a body (not labeled as such, but often with headings and subheadings), a Conclusions section, and a Literature Cited or References section. Most also include an Abstract, or Summary. For academic assignments, your instructor may have specific requirements about the format and components of the paper. If not, a good strategy is to model your own paper after a published review in your general subject area.

Many journals, such as *Biological Bulletin* and *American Midland Naturalist*, publish review articles in addition to research papers. Other journals specialize entirely in reviews; these include *Botanical Review, Physiological Reviews, Psychological Review, Quarterly Review of Biology,* and various annually published volumes of reviews, such as *Annual Review of Ecology and Systematics, Annual Review of Microbiology,* and *Annual Review of Genetics.*

Before you plunge into the first draft, you will need a tentative plan. This can range all the way from a rough sketch of the order of topics to a formal, detailed outline. Some people feel hemmed in by outlines and prefer to do much of their organizing as they work out the first draft. (These are the writers who, if ever asked to produce an outline, do so *after* they have written at least one draft of the paper.) Other people depend on a fixed plan right from the beginning to organize their thoughts. There is no one correct way to plan a review paper. You must decide what kind of organizational method suits your own writing style.

Many lengthy published reviews include a Table of Contents, which serves as a kind of outline of the paper. Shorter reviews, including course-based assignments, usually can do without this feature as long as they are well organized. Check with your instructor about his or her preferences. Here is a sample Table of Contents for an animal behavior paper:

```
THE ADAPTIVE SIGNIFICANCE OF ALARM CALLS IN MAMMALS
    Abstract . . . . . . . . . . . . . . . . . . . . 2
    Introduction . . . . . . . . . . . . . . . . . . 3
    Alarm Calls in Selected Mammals. . . . . . . . . 5
          Ground Squirrels. . . . . . . . . . . . . . 5
          White-Tailed Deer . . . . . . . . . . . . . 8
          Kloss's Gibbon . . . . . . . . . . . . . . 12
    Hypotheses for the Evolution of Alarm Calls . 15
    Conclusions . . . . . . . . . . . . . . . . . . 18
    Literature Cited. . . . . . . . . . . . . . . . 20
```

Note the use of headings and subheadings in this paper to give order to an otherwise unwieldy amount of material. Subdividing the text may also make the paper easier to write because you can tackle one chunk of material at a time. See page 203 for how to compose and format headings in your text.

■ Include an Abstract, if appropriate.

A published review typically starts with an Abstract stating the purpose and scope of the paper, listing major points about the topic, and summarizing important conclusions. As in a research paper, the Abstract is placed after the title and author, before the main text of the paper. Some journals require a Summary at the end of the paper instead of an Abstract at the beginning.

Beginning writers often confuse the Abstract of a review paper with its Introduction. Consider the following rough draft, for example:

```
Canine Hip Dysplasia (CHD) is a multifactorial,
polygenic disease that affects various breeds of
domestic dogs. A congenital illness, CHD pro-
gresses in stages, causing aberrant remodeling
of the coxofemoral joint. In this paper I pro-
vide an overview of the causes and treatments of
CHD. First I define CHD, discuss major symptoms,
and compare different diagnostic methods. Next,
I discuss the genetic bases of CHD, as well as
important risk factors. This review also summa-
rizes various treatment options for CHD, includ-
ing drug therapy, surgery, and the regulation of
diet and exercise. Finally, the relevance of se-
lective breeding is discussed.
```

This paragraph focuses more on the writer's plans than on what the paper actually says. For this reason, it is more similar to an Introduction than to an Abstract. By contrast, the following paragraph does a better job summarizing the aims and scope of the paper as well as its major points:

```
This paper reviews the causes, diagnosis, and
treatment of Canine Hip Dysplasia (CHD), a mul-
tifactorial, polygenic disease that affects
various breeds of domestic dogs. A congenital
illness, CHD progresses in stages, causing aber-
rant remodeling of the coxofemoral joint. CHD is
typically detected through phenotypic expression
and pedigree analysis and is most common in
large dogs. Although CHD can be diagnosed clini-
cally, radiographic analysis is more accurate.
```

```
Treatment varies depending on severity of the
condition: drug therapy relieves pain in mild
cases, but severe cases require surgery. Preven-
tative measures such as diet and exercise regu-
lation may delay onset or even reduce phenotypic
expression. Current research raises important
questions about the roles of breeders and ortho-
pedic registries in controlling the incidence of
CHD through selective breeding.
```

Note that the writer concludes with a more specific point about the broader relevance of the topic.

■ Introduce the subject, explain your rationale, and state your central question, objectives, or thesis.

These three tasks need to be accomplished in the Introduction. As in a research paper, an effective strategy is to start with broad statements, explanations, and definitions that orient and educate the reader. Then work down to more specific issues. Why is this subject important? What approach have you taken? Will you be giving a comprehensive summary or one that is more limited? End the Introduction with a clear statement of the question you will address or the main point you wish to convey to the reader. Your objectives may be fairly specific — for example, to show that certain kinds of early childhood experiences predispose adolescent girls toward anorexia. Or you may wish to assess the current state of research on a particular problem, for instance, current treatment methods for AIDS, with the aim of making predictions about the next decade.

The length of the Introduction depends on your subject and the kind of coverage you plan to give it. An 8- to 10-page review might need only a single succinct paragraph to introduce it. A longer paper might require a whole page. Generally, one or two paragraphs are appropriate. Readers will soon become confused if you do not tell them your objectives fairly early in the paper.

Here is the Introduction from the first draft of a term paper for a molecular biology course. The numbers in parentheses refer to sources in the Literature Cited section (see discussion in Chapter 6). How might this student paragraph be improved?

```
Many different articles were read about the mol-
ecular genetics of human growth hormone. This
paper will focus especially on hGH deficiencies.
```

```
Human growth hormone (hGH) is a polypeptide hor-
mone, produced from within a gene cluster on
chromosome 17, that controls much of the physi-
cal growth of the infant and child (1,2). Since
time is limited, this paper cannot cover all
possible aspects of hGH, so a narrower approach
has been taken.
```

In this Introduction, two sentences, the first and the last, say nothing essential; the reader naturally assumes you have read about your chosen topic, and it is your responsibility as the writer to narrow the focus by stating exactly what that focus is. Remember that the Introduction gives you your first chance (perhaps your only chance) to interest the reader. Obviously your instructor must read the paper whether he or she wants to or not, but if your start is forceful and interesting, the paper will have a much better effect.

A second problem with the paragraph is that we don't get a clear sense of the writer's *specific* purpose or rationale. We know that the paper will focus mainly on hGH deficiencies, but we do not know how or why. If the writer fails to portray the subject as important or intriguing, it is difficult for the reader to feel it is.

Here is the second-draft version of the Introduction. Notice how the writer has omitted the unnecessary sentences and filled in the gaps by expanding the more important parts of the original. The revised Introduction conveys a clearer idea of what this paper is about and why this subject is interesting.

```
Human growth hormone (hGH) is a polypeptide hor-
mone, produced from within a gene cluster on
chromosome 17, that controls much of the physi-
cal growth of the infant and child (1,2). Defi-
ciency of hGH, a heritable disorder, can result
in infantile dwarfism and retardation (3,4,5).
New research methods, including recombinant DNA
technology, have made it possible to determine
the molecular basis of such deficiencies. In
this paper, I will summarize current knowledge
of the molecular genetics of hGH and suggest
ways in which continued research may help physi-
cians treat infants with a deficiency of this
hormone.
```

■ Build a focused discussion.

Many student review papers are little more than summaries, boring ones in which the writer has retreated from the reader's sight. Just to regurgitate the contents of a series of papers one by one is not enough. You need to *relate* this material to your principal objectives. Present your information selectively, and use it to support or illustrate the statements you wish to make. A good review interprets the literature from the writer's own informed perspective and gives the reader a sense of integration, development, and focus.

The following paragraph is from a paper on the adaptive value of cannibalism in animals. Notice how the author uses examples from the literature to illustrate and develop the generalization in the first sentence.

> Following the reasoning of West Eberhard (1975),
> we may predict that cannibalism may be more
> likely when the potential victims are unusually
> vulnerable and easily obtained as food. Such in-
> dividuals are, in fact, the predominant victims
> of cannibalism in many species. For example,
> most cannibalism in Tribolium is performed by
> larvae and adults on the defenseless eggs and
> pupae (Mertz and Davies 1968). Pupae are also
> eaten by larvae in caddisflies (Gallepp 1974),
> and injured or weak immature stages are devoured
> by older nestmates in many species of social
> ants (Wilson 1971). In crows (Corvus corone),
> cannibalism of eggs and nestlings by intruding
> adults is more frequent when the parents are ab-
> sent from the nest, leaving their young more
> vulnerable to attack (Yom-Tov 1974).

The next passage is from a paper for a plant pathology course. It discusses host-specific toxins, substances produced by pathogenic (disease-producing) fungi that attack certain plants. Notice that the author presents selected information from the literature to critically examine a particular hypothesis. The writing conveys authority and a thorough familiarity with the material.

> Changes in the permeability of host cell mem-
> branes after being exposed to toxins suggest
> that these substances may bind to a receptor
> site in the cell membranes of susceptible hosts.

Strobel (1974) claims to have found such a re-
ceptor site in the membranes of sugarcane cells
treated with helminthosporoside, a toxin isolated from the pathogenic fungus Helminthospo-
rium sacchari. The site contains a protein of
molecular weight 48,000 daltons that specifically binds the toxin. Strobel has proposed that
the binding of the toxin stimulates the activity
of an adjacent membrane-bound enzyme, potassium-
magnesium ATPase, which maintains ion balance
across cell membranes. Such stimulation could
disrupt the membrane's selectivity to ions, resulting in the characteristic symptom of elec-
trolyte leakage.

Several authors have criticized Strobel's
methodology and interpretations. For example,
Wheeler (1976) doubts that the preparation used
for helminthosporoside assay and structure determination was sufficiently free from impuri-
ties, and he argues that some experiments lacked
sufficient replication. Others have said that
the toxin-binding data were not graphed correctly and that regraphing them suggests the
binding activity is, at best, weak (see, for example, Hanchey and Wheeler 1979). These and
other criticisms have been reviewed by Yoder
(1980).

■ Document your paper thoroughly.

Whenever you refer to another author's work or ideas, cite your
sources using conventional methods of literature citation (see Chapter 6).
In a review paper (as in the Introduction and Discussion of a research
paper), you need to cite references repeatedly. Look at the following sentence from a review by Hepler and Wayne (1985, p. 412), in which sources
are cited by numbers placed in parentheses:

> Red light triggers a large array of physiological and developmental
> events that require Ca^{2+}, including chloroplast rotation in *Mougeo-
> tia* (55,78,247–249), spore germination (254–256) and cell

expansion (37) in *Onoclea*, leaflet closure in *Mimosa* (22,23,237), root tip adhesion in *Phaseolus* (229,279), peroxidase secretion in *Spinacia* (113,164,165), membrane depolarization in *Nitella* (261), as well as activation of NAD kinase (1,218,232) and inhibition of mitochondrial ATPase (212).

Using many citations may seem strange at first. You may feel they impede the flow of your writing. However, readers of scientific papers are accustomed to such interruptions, and you will get used to them, too. Remember that literature citations serve an important function: they tell readers where to find additional information. Careful documentation also reflects the thoroughness of your literature search and your honesty in acknowledging the sources of your material.

■ Use quoted material sparingly.

Many beginning writers, unsure of their own voices or uncomfortable with the material, tend to fall back on quotations to get them through rough spots in the paper. Sometimes they construct whole paragraphs around a series of quotations from different authors, stringing these together with a few scattered phrases or sentences of their own. The text thus becomes a collection of other people's words:

> Studies of the Baltimore butterfly (*Euphydryas phaeton*) showed that "larvae occupied communal nests of various sizes" and "commonly cannibalized unhatched eggs in the same colony" (Ketchum 2003). "In both field and laboratory tests, there was a higher incidence of cannibalism by larvae occupying large colonies" (Ketchum 2003). Also, "cannibalistic acts occurred at a higher frequency under conditions of food shortage," when the larval foodplant, turtlehead, "was in short supply or extensively defoliated" (Joy 2000).

In such a passage the reader loses track of the writer, and the writer loses authority. The quotations do not enhance the text — they detract from it, suggesting that the writer hasn't come to terms with the material and is either too inexperienced or too lazy to use his or her own words. Biological authors rarely use quoted material, relying instead on careful, concise paraphrases or summaries; you should do the same. For example, the quotation-ridden passage above can be rephrased and condensed in the author's own words:

Studies of the Baltimore butterfly (*Euphydryas*
phaeton) showed that cannibalism of eggs by
the communal larvae was more frequent in large
colonies than in small ones (Ketchum 2003). Cannibalism also increased when the larval food-
plant, turtlehead, was scarce (Joy 2000).

When *is* it appropriate to use the exact wording of an author? Occasionally you may wish to include a quotation to establish or emphasize an
important point or state a precise definition. Sometimes you may feel that
an author's exact words are indispensable in conveying a particular viewpoint or idea:

> Bem (1981, p. 255) defines a *schema* as a "cognitive structure, a net-
> work of associations that organizes and guides an individual's per-
> ception."

> To Eiseley (1961), Darwin was "the man who saw the wrinkled
> hide of a disintegrating planet, glyptodonts and men, all equally
> flowing down the direction of time's arrow; he was a master artist
> and he entered sympathetically into life" (p. 351).

> In his controversial book *The Naked Ape*, Morris (1967, p. 211)
> concludes that the survival of the human species depends on an
> increasing awareness of our biological heritage:

>> We must somehow improve in quality rather than in sheer
>> quantity. If we do this, we can continue to progress technolog-
>> ically in a dramatic and exciting way without denying our
>> evolutionary inheritance. If we do not, then our suppressed
>> biological urges will build up and up until the dam bursts and
>> the whole of our elaborate existence is swept away in the
>> flood.

When you do use quoted material, cite the source in the text using either the name-year or number method (see Chapter 6). Some biological
authors include the page number of the book or article in the citation;
others do not, depending on the documentation style followed. Be sure to
reproduce the quoted material *exactly*. Introduce and punctuate quotations
properly, using the following rules.

1. Do not substitute single quotation marks (' ') for double ones (" ").
Single quotation marks are restricted to quotations within other quotations. (In British usage, single quotation marks are used first; double quota-
tion marks are used to set off quotations within quotations.)

2. A short quotation can be integrated into the text. Make sure it fits
in grammatically with the rest of the sentence.

Dawkins (1976, p. 206) suggests that memes, as units of cultural
transmission, can replicate themselves by "leaping from brain to

brain via a process which, in the broad sense, can be called imitation."

For a quotation longer than four typed lines in your paper, omit the quotation marks and indent each line of the passage five spaces from the left margin, but keep it double-spaced. A colon is often used to introduce the passage.

3. Place periods and commas *inside* the quotation marks; semicolons and colons go *outside*. (In British usage, periods and commas are placed outside quotation marks, as are semicolons and colons.)

> Jones (1987, p. 2) calls Davidson's explanation "the most exciting model of this century."

> Davidson's explanation is "the most exciting model of this century," according to Jones (1987, p. 2).

> Jones (1987, p. 2) considers Davidson's explanation "the most exciting model of this century"; unlike previous models, it gives rise to far-ranging predictions about how major evolutionary changes occur.

If exclamation marks, question marks, and dashes are part of the quoted material, put them inside the quotation marks. If they are part of your own sentence, put them outside.

4. If you need to interrupt a quotation to omit one or more words, indicate the omission by an ellipsis (three periods, separated by spaces). If the ellipsis falls at the end of a sentence, put the period followed by a space before the ellipsis.

> Morris (1976, p. 211) concludes that humans must "somehow improve in quality. . . . If we do this, we can continue to progress technologically in a dramatic and exciting way without denying our evolutionary inheritance."

5. If you insert clarifying or explanatory material into a quotation, put such material in brackets [ ].

> Morris (1976, p. 211) concludes: "If we do this [improve in quality], we can continue to progress technologically in a dramatic and exciting way without denying our evolutionary heritage."

■ End with general conclusions.

Despite the importance of an effective Conclusions section, it gets the least attention from many beginning writers of a review paper. By the time they reach this point, most writers are worn out and feel they have nothing more to say. They are so anxious to be done with the paper that they make a hasty retreat, summing up the topic in a sentence or two and then coming to a weary stop.

Thus the role of host-specific toxins in plant
disease is a complex topic. There are still many
questions for scientists in the future to answer.

The following student paragraph, however, summarizes the major
points in the paper and adds a final perspective:

In conclusion, current evidence shows that resistance or susceptibility is an inborn trait of
a plant. Some plant pathogenic fungi can produce
toxins that, by disturbing host physiological
functions, affect the onset and development of
plant disease. However, it is not clear whether
these substances have a primary role in determining resistance or susceptibility. Perhaps
some other mechanism may determine specificity,
and toxins may exert their damaging effects sec-
ondarily. Future research on host-specific toxins will help to answer this question.

As shown above, a Conclusions section consolidates and strengthens
the relationships, patterns, and arguments you have been building in the
body of the paper. In addition, a good Conclusion should do more than
merely summarize; it must also *conclude* something. Assuming that you have
adequately addressed the topic, now you must answer such questions as "So
what?" or "What next?" What is the significance of everything you have
just told the reader? What conflicts still need to be resolved? What research
must still be done? What might we expect to happen in the future? Notice
how the student writer of the following passage has begun to consider
such issues in her review of intervertebral disk herniation (IVD) in the
dog:

Within the past decade, some of the controversy
regarding the use of fenestration, dorsal
laminectomy, and hemilaminectomy for treating
IVD herniations has been resolved. Evidence now
suggests that routine fenestration of the affected disc, coupled with the selective use of
prophylactic fenestration, may be the safest and
most beneficial means of lowering recurrence
rates for IVD disease. However, far more research is needed to identify and characterize

```
the risk factors that predispose a patient to
multiple, and in particular adjacent, disc ex-
trusions. Accordingly, study designs will need
to become more consistent both in methodology
and in definition of terms....
```

Do not, however, introduce *new* information that really belongs in the body of the paper. Avoid complex questions or issues that you can't fully address. You do not want to throw readers off course or leave them hanging in midair. On the contrary, you want to tie up loose ends and finish with a satisfying sense of closure.

There is no set length for a Conclusions section. In published reviews, it varies from a single concise paragraph to a page or more of text. The length depends on the topic, the author's aims and depth of coverage, and any length restrictions on the paper. Plan ahead as you are drafting the manuscript so that you have enough space, time, and energy to end your paper effectively.

CHECKLIST FOR REVIEW PAPERS

Title

- Is your title specific and informative?
- Does it contain relevant keywords?

Abstract

- Does your Abstract state the aims and scope of your paper?
- Does it summarize major research on your topic?
- Is your Abstract concise?

Introduction

- Does your Introduction present the scientific issue and give a rationale for your review?
- Do you state the paper's central question or objectives?

Body of the paper

- Do you synthesize and interpret the literature, rather than merely summarize?
- Are your paragraphs well organized, developing main points by means of details and specific examples?
- Are headings and subheadings parallel, focused, and concise?

- Do you rely on your own wording, using quotations rarely if at all?
- Have you cited all important sources?

Conclusions

- Is your Conclusions section more than just a summary?
- Do you discuss controversies in the literature, the need for new research, and broader implications of your topic?

Acknowledgments

- Have you thanked all those who helped with the research or writing of the paper?

References/Literature Cited

- Are all sources cited in the paper also listed in your References?
- Does your References section contain any sources not cited in the paper?
- Have you documented your references using a conventional format consistently and meticulously?
- Is the information about every source accurate and complete?

SAMPLE REVIEW PAPER

The following excerpt is from a student review paper for a physiology course. References are listed using the CSE citation-sequence documentation style (see Chapter 6).

Include title,
author, course,
and date on
title page for
student papers.
Do not number
title page, but
consider it
page 1.

The Role of Hypothermia and the Diving
Reflex in Survival of Near-Drowning
Accidents

Brian Martin

Biology 281
April 17, 200-

Hypothermia and the Diving Reflex 2

ABSTRACT

This paper reviews the contributions of hypothermia and the mammalian diving reflex (MDR) to human survival of cold-water immersion incidents. The effect of the victim's age on these processes is also examined. A major protective role of hypothermia comes from a reduced metabolic rate and thus lowered oxygen consumption by body tissues. Although hypothermia may produce fatal cardiac arrhythmias such as ventricular fibrillation, it is also associated with bradycardia and peripheral vasoconstriction, both of which enhance oxygen supply to the heart and brain. The MDR also results in bradycardia and reduced peripheral blood flow, as well as laryngospasm, which protects victims against rapid inhalation of water. Studies of drowning and near-drowning accidents involving children and adults suggest that victim survival depends on the presence of both hypothermia and the MDR, as neither alone can provide adequate cerebral protection during long periods of hypoxia. Future lines of research are suggested and related to improved patient care.

INTRODUCTION

Drowning and near-drowning incidents are leading causes of mortality and morbidity in both children [1,2] and adults [3,4]. Over the past 30 years, there has been

State aims and scope; concisely summarize major points.

Introduce topic; give paper's aims and scope. CSE citation-sequence format is used to cite sources, which are numbered in order of mention in the text.

Hypothermia and the Diving Reflex 3

considerable interest in cold-water immersion incidents, particularly the
reasons for the survival of some victims under seemingly fatal conditions.
Giebrecht [5], for example, has reviewed
some of this literature. Research suggests that both hypothermia and a "mammalian diving reflex" (MDR) may account
for survival in many near-drowning
episodes [6]. However, the extent to which
these two processes interact is not fully
understood. Controversy also exists regarding the effect of the victim's age on
the physiological responses to cold-water
immersion. In this paper, I provide an
overview of recent research on the protective value of hypothermia and the MDR
in cold-water immersions. I also examine
hypotheses concerning the effects of age
on these processes and conclude with suggestions about future lines of research
that may lead to improved patient care.

*Hypoxia during drowning and near-drowning
incidents*

The major physiological problem fac-
ing drowning victims is hypoxia, or lack
of adequate oxygen perfusion to body
cells [1,7]. Hypoxia results in damage to
many organs, including the heart, lungs,
kidneys, liver, and intestines [7]. Generally, the length of time the body has
been deprived of oxygen is closely related to patient prognosis. Only 6-7 s of

In body of paper, summarize and integrate data from primary sources. Use headings, if desired, to organize text.

Hypothermia and the Diving Reflex 4

hypoxia may cause unconsciousness; if
hypoxia lasts longer than 5 min at relatively warm temperatures, death or irre-
versible brain damage may result [8]. However, some victims of immersion in cold
water have survived after periods of
oxygen deprivation lasting up to two
hours [7]. . . .
[The student goes on to highlight the major controversies
and to add interpretation and analysis.]

Use original
number as-
signed to refer-
ence when cit-
ing it again.

CONCLUSIONS

Recent research on cold-water immersion incidents has provided a more complete understanding of the physiological
processes occurring during drowning and
near-drowning accidents. Our current understanding is that the cooperative effect of the MDR and hypothermia plays a
critical role in patient survival during
a cold-water immersion incident [6]. However, the relationship between the two
processes is still unclear. Because it is
impossible to provide an exact reproduction of a particular drowning incident
within the laboratory, research is ham-
pered by the lack of complete details
surrounding drowning incidents. Consequently, it is difficult for comparisons to be drawn between published case
studies.

Conclude with
suggestions for
future research,
links to broader
issues.

More complete and accurate documentation of cold-water immersion incidents — including time of submersion;
time of recovery; and a profile of the

Hypothermia and the Diving Reflex 5

victim including age, sex, physical condition — will facilitate easier comparison of individual situations and lead to a more complete knowledge of the processes affecting long-term survival rates for drowning victims. Once we have a clearer understanding of the relationship between hypothermia and the MDR — and of the effect of such factors as the age of the victim — physicians and rescue personnel can take steps to improve patient care both at the scene and in the hospital.

ACKNOWLEDGMENTS

With their permission, acknowledge people who assisted with the paper.

I would like to thank V. McMillan and D. Huerta for their support and suggestions throughout the production of this paper. I am also grateful to my classmates in Biology 281 for their thoughtful comments during writing workshops. Finally, I thank Colgate University's interlibrary loan staff for help securing the sources needed for this review.

LITERATURE CITED

Include all references cited. Numbers correspond to order in which references were first mentioned in the text. For journal articles, CSE format calls for author's last name, followed by initials, then article title, abbreviated journal title, publication date, volume and issue numbers, and inclusive page numbers.

1. Kallas HJ, O'Rourke PP. Drowning and immersion injuries in children. Curr Opin Pediatr. 1993;5(3):295-302.
2. Wollenek G, Honarwar N, Golej J, Marx M. Cold water submersion and cardiac arrest in treatment of severe hypothermia with cardiopulmonary bypass. Resuscitation. 2002;52(3): 255-263.

Hypothermia and the Diving Reflex 6

3. Keatinge WR. Accidental immersion hypothermia and drowning. Practitioner. 1977;219(1310):183-187.

4. Mehta SR, Srinivasan KV, Bindra MS, Kumar MR, Lahiri AK. Near drowning in cold water. J Assoc Physicians India. 2000;48(7):674-676.

5. Giesbrecht GG. Cold stress, near drowning and accidental hypothermia: a review. Aviat Space Environ Med. 2000;71(7):733-752.

6. Gooden BA. Why some people do not drown — hypothermia versus the diving response. Med J Aust. 1992;157 (9):629-632.

7. Biggart MJ, Bohn DJ. Effect of hypothermia and cardiac arrest on outcome of near-drowning accidents in children. J Pediatr. 1990;117(2 Pt 1):179-183.

8. Gooden BA. Drowning and the diving reflex in man. Med J Aust. 1972;2(11): 583-587.

9. Bierens JJ, van der Velde EA. Submersion in the Netherlands: prognostic indicators and the results of resuscitation. Ann Emerg Med.1990;19(12): 1390-1395.

10. Ramey CA, Ramey DN, Hayward JS. Dive response of children in relation to cold-water near drowning. J Appl Physiol. 1987;62(2):665-668.

Hypothermia and the Diving Reflex 7

11. Wong KC. Physiology and pharmacology of hypothermia. West J Med. 1983;138(2):227-232.

12. Kemp AM, Sibert JR. Outcome in children who nearly drown — a British Isles study. Br Med J. 1991;302(6782):931-933.

13. Conn AW, Edmonds JF, Barker GA. Cerebral resuscitation in near drowning. Pediat Clin North Am. 1979;26(3):691-701.

CHAPTER 6

Documenting the Paper

CITING SOURCES IN THE TEXT

■ Acknowledge the source of all material that is not your own.

The text of a biological paper usually contains numerous *literature citations*, or references to the published studies of other authors. This is because scientists rarely work in a vacuum: hypotheses are developed, tested, and evaluated in the context of what other scientists have written and discovered. Most readers of biological literature are interested not only in the specific study being described but also in the bigger picture—how particular findings contribute to our current understanding of an important scientific problem. Thus, careful *documentation*, or acknowledgment of the work of others, is essential to good scientific writing.

Biologists also need to provide literature citations because, like other writers, they have an ethical and legal obligation to give credit to others for material that is not their own. Such material includes not only direct quotations but also findings or ideas that stem from the work of someone else.

■ Cite sources using a conventional format accepted by biologists.

Citation formats vary from one academic field to another. Unlike writers in the humanities, for example, biologists rarely use footnotes or endnotes to acknowledge sources. Instead, they usually insert citations directly in the text, either by giving the last name of the author(s) and publication date or by referring to each source by a number. Complete bibliographic information is listed for each cited reference at the end of the paper in a section called "Literature Cited," "References," or "Cited References." Each citation format—name-year or number method—has its advantages and disadvantages, and certain subdisciplines of biology (and each biological journal) tend to favor one method or the other.

A major advantage of the name-year system is that it is more informative. Many readers of biological literature are familiar with the major works in a particular field and will recognize well-known papers that are cited by the writer. In addition, the date of a particular study may be important to mention if its findings have been confirmed (or challenged) by later research.

Citing sources by number, however, conveys no additional information. Readers who want to know who did the study, and when, must track down this reference in the Literature Cited section, unless the writer has supplied additional information in the text. Also, simple typographical errors—for example, typing the number 13 instead of 14—can easily introduce inaccuracies. If a citation is added or deleted at the last minute, the writer must spend time renumbering all the references (although some word processors automatically renumber references). However, some people find the number method less cumbersome and distracting, especially if the text contains a large number of citations. Therefore, this format is often used for review papers (see Chapter 5).

Which documentation method should you use? If you are preparing a paper for publication, you should follow the guidelines of the particular journal you hope will accept the paper. These guidelines, usually titled "Instructions to Authors," are published regularly in print issues of the journal and are usually available on the journal's Web site, as well. If you are unable to locate the Instructions to Authors for a particular journal, ask a librarian for help. The Instructions can be requested as an interlibrary loan item just as you would request a journal article. Submission requirements vary widely among scientific journals, and they can be very detailed, including not only documentation style but also many other aspects of the manuscript. If you want your paper to be given prompt and serious consideration, you need to follow these guidelines carefully, down to the last detail.

For an academic assignment, first ask your instructor which documentation method he or she prefers. If you are free to choose, then adopt one of the formats discussed below or use the format of a published paper or a representative journal in your field of study. Whatever method you select, be consistent and meticulous in adhering to the conventions. Such rules, even if they seem arbitrary, make the reporting of references an orderly activity, minimizing confusion for both writers and readers.

■ CSE Style

The following citation guidelines are based on recommendations of the Council of Science Editors (CSE <http://www.councilscience editors.org>). The CSE was formerly called the Council of Biology Editors (CBE), so you may already be familiar with "CBE style" as described in *Scientific Style and Format: The CBE Manual for Authors, Editors, and Publishers* (1994). CSE guidelines, in turn, are based largely on the *National Library of Medicine Recommended Formats for Bibliographic Citation* (1991), and they are followed closely, or with some variations, by many journals in biology, geology, medicine, chemistry, mathematics, and physics. The following summary of CSE format reflects current guidelines as outlined in the current (6th) edition of the manual, along with some changes likely to appear in the 7th edition. The CSE now recognizes three documentation options: the *name-year* format plus two variations of the number format—the *citation-sequence* and the *citation-name* systems.

CSE name-year system

Cite each reference by giving the name(s) of the author(s) followed by the year in which the material was published.

WORK BY ONE AUTHOR

For each citation, use parentheses to enclose the name and the date. The name is followed by a space, but no comma:

```
The most recent study of coral bleaching in this
region (Scipione 1995) suggests that . . .
```

If the author's name appears as part of the sentence, put just the date in parentheses:

```
In Chandler's (1993) study of marine planktonic
ciliates . . .
```

```
Black-horned locusts were first reported in
Maryland by Rampolla (1980).
```

Because a name-year citation is considered part of the text, the period should always *follow* a citation that appears at the end of your sentence:

INCORRECT	`The presence of hairs on a leaf can increase light reflectance from the leaf surface. (Cunningham 1966)`
INCORRECT	`. . . leaf surface (Cunningham 1966.)`
CORRECT	`. . . leaf surface (Cunningham 1966).`

WORK BY TWO AUTHORS

Put the senior author's name first. The senior author is the one whose name appears first after the title. It is generally assumed that he or she has had major responsibility for writing the paper.

```
Other researchers (Silsby and Dunkle 1981) have
suggested a different method of chemical analy-
sis.
```

```
Hagihara and Inoue (1993) found that . . .
```

WORK BY THREE OR MORE AUTHORS

Cite the senior author's name followed by "et al." or "and others" and then the date:

```
White-lined bark beetles are attracted to the
odor of rotting wood (Zorn and others 1992).
```

```
Henry and others (2000) have suggested that . . .
```

Note the use of *have* (not *has*) in the example above. The subject (Henry and others) is plural and therefore must have a plural verb.

TWO OR MORE WORKS BY THE SAME AUTHOR

Give the author's name followed by the dates of the works in chronological order. Use a comma followed by a space to separate the dates:

```
The circulatory system of this species has been
described in detail by Wylie (1978, 1980, 1983).
```

```
Recent work on reaggregation of sponge cells
(Zangrilli 1994, 1995) has shown that . . .
```

TWO OR MORE WORKS BY THE SAME AUTHOR IN THE SAME YEAR

Use letters (a, b, and so on) to differentiate between references. Retain this designation when listing the references in the Literature Cited section.

```
Rush (2000a, 2000b) postulated . . .
```

TWO OR MORE WORKS BY DIFFERENT AUTHORS

To cite more than one reference within the same parentheses, list them in chronological order, with the earliest first, and separate them using a semicolon followed by a space. To separate works by the *same* author, use a comma:

```
Many different models have been proposed to ac-
count for this phenomenon (Watson 1977, 1979;
Kim 1988; Cox 1992).
```

If two or more sources are by different authors but published in the same year, put them in alphabetical order. In this example, Carter precedes Kim:

```
Many different models have been proposed to ac-
count for this phenomenon (Watson 1977, 1979;
Carter 1988; Kim 1988; Cox 1992).
```

AUTHORS WITH THE SAME LAST NAME

Occasionally two works published in the same year will have authors with the same last name. In this case, give the initials of each author in the citation:

```
Similar findings were reported for Colorado pop-
ulations of this species (Rider R 1990; Rider WA
1990).
```

WORK WITH AN ORGANIZATION AS AUTHOR

Give the name of the organization or group, followed by the year of publication.

```
Guidelines for evaluating dragonfly habitats as
potential conservation areas have recently been
published (International Committee for Preserva-
tion of Odonata 2004).
```

You may abbreviate organization names if they are long or if you need to cite them frequently:

```
Guidelines for evaluating dragonfly habitats as
potential conservation areas have recently been
published (ICPO 2004).
```

In the Literature Cited section, include the abbreviation at the beginning of this entry:

```
[ICPO] International Committee for Preservation
of Odonata. 2004.
```

WORK WITH AN UNKNOWN AUTHOR

The CSE does not advise use of the word "Anonymous" if no author (person or organization) is named in a work. Instead, use the first few words of the title in place of the author's name, and add an ellipsis before the date:

```
This species has been reported in both Madison
and Oneida Counties of New York (Common dragon-
flies . . . 1999).
```

ARTICLE OR CHAPTER IN AN EDITED VOLUME

To cite a reference in an edited collection of works by different authors, do not use the editor's name. Just give the name of the author(s) of the particular article or chapter, along with the year the edited collection was published. For example, suppose that one of your sources is "Sexual Selection and Mate Choice," by Timothy R. Halliday, which appears on pages 180–213 in the book *Behavioural Ecology: An Evolutionary Approach*, edited by J. R. Krebs and N. B. Davies, and published in 1978. Your in-text citation would be (Halliday 1978) and *not* (Krebs and Davies 1978).

A SPECIFIC PART OF A WORK

If you need to refer to a particular table, figure, or other portion of a source, it is acceptable to include this information within the parentheses, as shown:

```
Simmons and Grimaldi (1999, Table 4) found
that . . .

Similar data were obtained by Linsley (2004,
Fig. 2).
```

Note the use of a comma, followed by a space, after the date.

A DIRECT QUOTATION

Biological authors rarely use quoted material, relying instead on concise paraphrases and summaries when referring to other works (see p. 30). If a writer does include a direct quotation, the page number often appears as part of the citation:

```
Rith-Najarian (1998, p. 3) has written, ". . .
```

UNPUBLISHED MATERIAL

Occasionally you may need to cite a personal conversation with some one, an informal written communication, or the unpublished data of a colleague. E-mail correspondence—for example, with colleagues or with a known expert in a particular field—also falls into this category. Professional ethics dictate that you obtain the person's permission to include such material in your paper. For the citation, provide the last name and initials of the person who supplied the information, along with the date and the type of communication. Similarly, if you refer to your *own* unpublished data from a *different* study, give your name followed by a brief explanation of the material. Do not list unpublished sources in the Literature Cited section because they are inaccessible to your readers.

```
D. Craine (phone conversation, 2005) has sug-
gested that . . .

This species has also been found along the
shores of lakes and ponds (L Arnold, unpublished
data, 2000).
```

Professional biologists are often familiar with their colleagues' recent manuscripts and may refer to them in their own writing. When an article or book has been accepted for publication but has not yet been printed, it is regarded as "in press" or "forthcoming." The CSE prefers the term "forthcoming" because it can be used not only for printed materials but also for electronic sources. When a source is forthcoming, indicate this in the citation, along with the anticipated date of publication:

```
Wilson and Joseph (forthcoming, 2005) found
that . . .
```

If you are uncertain about the future publication date for a particular source, you may omit the date from the citation.

WORKS YOU HAVE NOT CONSULTED DIRECTLY

The CSE advises that such sources *not* be included as references. In other words, to avoid possible inaccuracies or misinterpretations, do not rely completely on another author's account of a particular source. Cite *only* those sources that you have consulted directly, so that readers can assume you have firsthand knowledge of all the works you discuss.

COURSE HANDOUTS OR LABORATORY MANUALS

For academic assignments, you may wish to cite an unpublished handout or lab manual that is specific to a particular course. One way to do this, if you are using the name-year citation system, is to give your instructor's name (as author) along with the year:

```
Sampling methods followed the procedure outlined
by Lynes (2005).
```

If there is no clear author of the lab manual (as in the case of a frequently offered course by a changing team of instructors), then give the title of the manual followed by the year:

```
Community similarity was determined using a mod-
ification of Morisita's Index (Ecology 204 Labo-
ratory Manual 2005).
```

Your instructor may have his or her own preferences about in-text citations of course materials.

CSE citation–sequence and citation–name systems

These documentation systems are different versions of the number
format. Here each source is assigned a number, which is then used to refer
to the source in the text of the paper. If the same source is cited more than
once, it retains the number originally assigned to it. In the *citation-sequence
system*, sources are numbered in the order in which you cite them in your
paper. They are then listed in numerical order in the References section. In
the *citation-name system*, sources are first alphabetized and *then* numbered.
These numbers are used for the in-text citations, and sources are listed alphabetically in the References.

The citation number often appears as a superscript, placed either at the
end of the sentence or within the sentence at the appropriate location:

```
Sap pressures in Douglas fir become more nega-
tive as relative humidity rises [1].

Recent field studies [12] have suggested that . . .
```

Note that a space is placed before the numeric citation, as shown above.
Some authors enclose the reference number in parentheses rather than
using a superscript: for example, (12):

Whichever version of the CSE number format you use, the following
guidelines apply:

If you cite two or more sources at the same point in the text, use commas (with no intervening spaces) to separate the numbers. If three or more
of your reference numbers are consecutive, use an en dash (or a hyphen) to
indicate this:

```
Plantago major is common in heavily trampled
areas, such as the edges of roads and side-
walks [2,3].

Cytological studies of development from dormant
gemmules [4,6-8] have shown that . . .
```

If the author or publication date of a particular study is important to
your discussion, add this information to the sentence:

```
Fuller ¹⁴, studying three species of caddisflies
in Montana, was the first to observe that . . .
```

```
This species was not listed in early floras of
New York; however, in 1985 it was reported in a
botanical survey of Chenango County ⁹.
```

Finally, note in the example above that when the number (superscript) appears at the end of a sentence, it is placed *before* the period. The CSE currently has no specific rules about the placement of a comma before or after a superscript.

■ APA Style

APA style is based on recommendations by the American Psychological Association <http://www.apa.org>. These guidelines are described at length in the 5th edition of the *Publication Manual of the American Psychological Association* (2001). APA format is used by many authors in psychology, nursing, and other behavioral and social science fields. Following are some examples of common in-text citations; see the APA manual for additional examples and details.

WORK BY ONE AUTHOR

For each citation, use parentheses to enclose the name and the date. Separate the name and date with a comma followed by a space:

```
These preliminary observations have been con-
firmed by recent laboratory experiments (Nathan,
2005).
```

Note that the citation appears *before* the period at the end of the sentence.

If the author's name appears as part of the sentence, put just the date in parentheses:

```
Nathan (2005) has suggested that . . .
```

WORK BY TWO AUTHORS

Put the senior author's name first. The senior author is the one whose name appears first after the title. Separate the two names with an ampersand (&) and put a comma between the second name and the date:

```
A subsequent study (Eurelle & Doyle, 2004) fo-
cused on . . .
```

If you use the authors' names in your sentence, substitute the word "and" for the ampersand:

```
Eurelle and Doyle (2004) found that . . .
```

WORK BY THREE, FOUR, OR FIVE AUTHORS

The first time you cite the source, give all the authors' names, separating the last two with an ampersand. If you use the names in your sentence, replace the ampersand with the word "and."

```
The most recent study of this process (Doyle,
Walsh, & Dowling, 2003). . .

Doyle, Walsh, and Dowling (2003) found that . . .
```

In subsequent citations of the same source, give just the first author's name along with the Latin abbreviation "et al." ("and others"):

```
Doyle et al. (2003) have proposed that . . .

. . . as proposed originally (Doyle et al., 2003).
```

Note the use of *have* (not *has*) in the first example above. The subject (Doyle et al.) is plural and therefore must have a plural verb.

WORK BY MORE THAN FIVE AUTHORS

In all citations, give just the senior (first) author's name, followed by "et al." and the year:

```
Averill et al. (2002) studied communication pat-
terns in . . .
```

TWO OR MORE WORKS BY THE SAME AUTHOR

Give the author's name followed by a comma and the dates of the works in chronological order, separated by commas:

```
Several studies (Hill, 1999, 2002) . . .
```

TWO OR MORE WORKS BY THE SAME AUTHOR IN THE SAME YEAR

Use letters (a, b, and so on) to differentiate between the works, repeating the year and using commas between the works:

```
Field studies by Wylie (2003a, 2003b) revealed
that . . .
```

TWO OR MORE WORKS BY DIFFERENT AUTHORS

To cite more than one author within the same parentheses, list them alphabetically and separate them using semicolons:

```
(Doyle et al., 2003; Fogg, 2004; Hill, 1998;
Opipari, 1999, 2000)
```

Note that Opipari's two works are separated by commas and listed chronologically.

Authors with the Same Last Name

If your references include different works by authors with the same last name, add the initials of the authors for clarity. The APA recommends that you follow this procedure even when the authors' works have different publication dates.

```
A. R. Johnson (2000) and C. Johnson (2002) both
studied . . .
```

```
These studies were later repeated in the labora-
tory (A. R. Johnson, 2000; C. Johnson, 2002).
```

Note the periods and spaces in the authors' initials.

Work with an Organization as Author

Give the name of the organization, followed by a comma and the date:

```
(Carolina Psychological Society, 2003)
```

If you need to cite the same organization later, you may abbreviate the organization the first time, in brackets, after the full name:

```
(Carolina Psychological Society [CPS], 2003)
```

Then use the abbreviation in subsequent citations:

```
(CPS, 2003)
```

Note that this abbreviation is *not* used in the References list. See the discussion on abbreviations on pages 141 and 154.

Work with No Identifiable Author or Anonymous Author

If there is no named author, give the title of the work or a shortened version of it, followed by a comma and the date. Put the title in quotation marks if the work is an article; put the title in italics if the source is a book:

```
("Coping Strategies in Adolescents," 1997)
```

```
(Stress Responses, 1988)
```

If the author is actually designated as "Anonymous," then use this word along with the date:

(Anonymous, 1973)

ARTICLE OR CHAPTER IN AN EDITED VOLUME

To cite a specific source in an edited collection of works, give the name of the author(s) of the article or chapter, along with the year the collection was published. For example, if you need to refer to an article by William A. Rider in a collection of articles edited by Elizabeth Hayes and published in 2002, then your in-text citation would be (Rider, 2002) and *not* (Hayes, 2002).

A SPECIFIC PART OF A WORK OR A DIRECT QUOTATION

If you need to refer to a particular table, figure, chapter, or other portion of a source, add this information to the citation. For a direct quotation, add the page number(s), preceded by the abbreviation "p." for "page" or "pp." for multiple "pages":

```
Surveys of 163 male college students (Ramsden,
2001, Table 3) showed that . . .
```

```
According to Oni (2001, p. 42), "No model has
yet been proposed that fully accounts for such a
wide range of human responses."
```

WORKS YOU HAVE NOT CONSULTED DIRECTLY

Avoid referring to sources you have not read yourself; the reader assumes that you have firsthand knowledge of all the works you have cited. Occasionally, however, you may need to cite an important source that is not accessible to you. If so, specify where you acquired your secondhand knowledge of this source:

```
Jasper (1969, as cited in Simon, 2004, p. 24)
suggested that cat owners and dog owners share
several key personality characteristics.
```

List *both* sources in your References.

UNPUBLISHED MATERIAL

To cite material from letters, e-mail, interviews, phone conversations, etc., give the last name and initials of the person with whom you communicated, along with the date (as exact as possible):

```
(E. J. Ketchum, personal communication, March
26, 2005)
```

Do not list this kind of source in your References, since readers cannot readily access it.

To refer to works that have been accepted for publication but have not yet been published, give the name followed by "in press" instead of the date. This source should appear in your reference list, along with the title and other information.

```
(Sydlik & Badgerow, in press)
```

COURSE HANDOUTS OR LABORATORY MANUALS

Refer to these sources by the author's (instructor's) name and the year, or by the title (italicized) and year if the authorship is not clear.

```
(Giles, 2005)
```

```
(Psychology 101 Laboratory Manual, 2005)
```

Your instructor may have his or her own preferences about citing such sources.

■ Put citations where they make the most sense.

Whichever documentation system you use, put each citation close to the information you wish to acknowledge. Do not automatically put cited material at the end of every sentence. For example, the following statement (using CSE style) is ambiguous:

```
Pollination of Linaria vulgaris has been studied
in both the field and the laboratory (Arnold
1962; Howard 1979).
```

Did Arnold do his studies in the field and Howard in the laboratory? Or Howard in the field and Arnold in the laboratory? Or both authors in both settings? Moving the first citation clarifies the situation:

```
Pollination of Linaria vulgaris has been studied
in both the field (Arnold 1962) and the labora-
tory (Howard 1979).
```

If you actually meant to say that both Arnold and Howard did both types of studies, then you are better off rewording the sentence:

```
Arnold (1962) and Howard (1979) have studied the
pollination of Linaria vulgaris in both the
field and the laboratory.
```

■ Do not cite sources for information regarded as common knowledge in a particular field.

Although it is not strictly wrong to do this, it is unnecessary. If certain material is well known and fundamental to a particular field, you need not cite sources. For example, you need not cite your textbook or other references to say that living organisms are composed of cells or that meiosis in higher animals gives rise to haploid gametes. Such material is general knowledge that is familiar to the audience of any biological paper. Similarly, any subdiscipline of biology has information that is regarded as elementary and basic by anyone working in that field.

How do you decide what is common knowledge and what is not? This ability comes with experience as you grow more familiar with a subject and the literature on it. For academic assignments, the background knowledge you share with your classmates will be your best guide. If in doubt, ask someone more experienced than yourself or cite the source anyway.

■ Use citations carefully.

Citations allow you to acknowledge the work of others; they also *inform* the reader. Do not pack your text with citations simply to demonstrate that you've found numerous sources on your topic. Consider the following example (CSE style):

```
Many studies have been made of the factors in-
fluencing variable mate-guarding in the dragon-
fly Plathemis lydia (Brubaker 1979; Darby 1980;
Kraly 1981; Jamieson, Pegg, & Joy 1983; Angell
1984; Napolin 1985; Hayes 1988; Gittens 1991).
```

In this passage, vague reference to a large number of works does little to enlighten the reader. Decide which of your many sources are *most* important and why. Refer to them in a meaningful way so that they promote a focused discussion:

```
Many studies have been made of the factors in-
fluencing variable mate-guarding in the dragon-
fly Plathemis lydia. Brubaker (1979) found
that . . . , whereas Darby (1980) showed
that . . . Experimental manipulations of male
density at breeding sites by Angell (1984) and
Hayes (1988) suggested that . . .
```

Watch out for unintentional sexism when you refer to authors. Do not automatically assume that every biologist is male; use the appropriate pronoun based on the author's first name:

```
Virkler (1995) investigated bacterial resistance
to the antibiotic chloramphenicol. She suggested
that . . .
```

■ Avoid unnecessary repetition of citations when the context is clear.

When using the CSE name-year system, you need not repeat the author's name if it already appears earlier in the sentence and you are not citing additional authors:

```
Huerta (1960) conducted the first laboratory
studies of this phenomenon and reported in later
works (1970, 1977) that . . .
```

When using either the CSE name-year or one of the number systems, you need not repeat the same citation in every sentence when you are discussing the work at some length. Again, let the need for clarity be your guide. In the following example, note that the writer has cited the source at the start of her paragraph and again, near the end, to remind the reader that she is still referring to the same study:

```
Nakhimovsky and Ochs (1994) studied interspe-
cific interactions between two species of libel-
lulid dragonflies (Libellula pulchella and
Plathemis lydia) at a small pond in upstate New
York. They found much overlap between the two
species in the use of shoreline areas for pair
formation, territoriality, mating, and ovi-
position. Interspecific interactions between
males seeking perch sites were relatively infre-
quent; however, male P. lydia often disrupted
oviposition attempts by L. pulchella females.
Nakhimovsky and Ochs (1994) suggest several
promising lines of research on sex recognition
in these two species, which are characterized by
strikingly similar females.
```

PREPARING THE LITERATURE CITED SECTION

■ Understand the difference between a Literature Cited section and a bibliography.

A bibliography contains all the sources mentioned in the text, along with additional references on the topic. The Literature Cited (or References or Cited References) section contains *only* the sources that have been *cited* (referred to) in the text. Even if you have acquired useful background knowledge by reading five articles and three books, do not list any of them in the Literature Cited section unless you have specifically mentioned them in your text. Bibliographies are not generally part of scientific papers.

■ Report sources completely and accurately.

After writing the main body of the paper, you may be tempted to race through the listing of references, assuming that no one looks at this part anyway. Do not underestimate the importance of a meticulously prepared Literature Cited section. Readers *do* look at this section. In their minds, the amount of attention you have given it reflects the care given to the rest of the paper. If your sources are reported sloppily, people may doubt your authority, integrity, and thoroughness as a researcher.

The Literature Cited section also serves as an important source of references for readers who want further information on the topic. You owe these people accuracy and completeness. Even some published papers contain mistakes or are missing information in the list of references. Few things are more frustrating for a researcher than to be told the wrong page numbers or journal volume for an article he or she needs to track down. Long *before* you sit down to prepare the Literature Cited section, become familiar with the kinds of bibliographic details you will need for each type of source (paper, book, article in an edited collection of articles, Web document, and so on). As you read and take notes, record all this information in a master list or on file cards. Doing so will save you time and avoid chaos later. See pages 29–32 for more on note-taking techniques.

■ Use a conventional format for listing your sources.

As with in-text citations, biological journals vary in the formats used for the Literature Cited or References section. In fact, you won't have to plunge too deeply into the primary literature to discover a bewildering array of variations from journal to journal. Prospective authors prepare the reference list by closely following the guidelines prescribed by the journal for which they are writing. For academic assignments, check with your instructor; otherwise, use one of the formats presented here or follow the

style of a recent paper on your topic or a journal in that
that the documentation format you use for your refere
the same one used for your in-text citations.

■ CSE Style

The following examples are based on CSE guidelines for listing references using either the name-year citation system or either of the two
number systems (citation-sequence and citation-name). The two number
systems use the same format in the references section. These examples
should help you with most of the references you're likely to encounter as
you begin biological research. See the CSE manual for information on citing specialized technical reports, government documents, audiovisual publications, newspaper articles, and other sources not covered here.

JOURNAL ARTICLE WITH SINGLE AUTHOR

NAME-YEAR SYSTEM	Straus DC. 1982. Protease production by *Streptococcus sanguis* associated with subacute bacterial endocarditis. Infect Immun. 38(3):1037-1045.
NUMBER SYSTEMS	36. Straus DC. Protease production by *Streptococcus sanguis* associated with subacute bacterial endocardi- tis. Infect Immun. 1982;38(3): 1037-1045.

In the CSE name-year format, the year of publication follows the author's name. After the year comes a period followed by the title of the
paper, the abbreviated journal title, a period, the volume of the journal, the
issue number (in parentheses), a colon, and the pages on which the paper
appears. Supply all digits for page numbers: 1037-1045, not 1037-45.

In the CSE number systems, the citation number (here, 36) used by
the writer to refer to this source in the text is listed first, followed by the
author's name and initials. Next is the title of the article, a period, and the
abbreviated journal title followed by a period. The year of publication
comes next, followed by a semicolon, the volume of the journal, issue
number (in parentheses), a colon, and then the pages on which the article
appears. There are no spaces between any of these components—from the
year to page numbers—in the reference listing.

Note that in both types of entries, the last name of the author is followed by his or her initials only; the initials are not separated by periods or
spaces. Only the first word of the article title is capitalized. The journal title
(here, *Infection and Immunity*) is abbreviated and capitalized but not underlined or italicized. (The Latin name of the bacterium *is* italicized, and the

...s is capitalized; see p. 197.) Journal titles are shortened using standard ...reviations; one-word journal titles are not abbreviated (e.g., *Science*). One way to find the conventional abbreviation for a journal is to look at a copy of the journal itself. You can also consult the MARC or machine-readable cataloging record in your library's online catalog or the Library of Congress online record for the journal. A comprehensive list of journal title abbreviations is available for downloading from the National Library of Medicine's Entrez database at <http://www.ncbi.nlm.nih.gov/entrez/getids_help.html#JournalLists>.

You will notice two differences between the name-year and number systems (the placement of the date and the presence or absence of the semicolon) in the following examples of other types of reference listings.

According to the most recent CSE guidelines, both the *volume* and the *issue* number should be provided for journal references. This is the case whether or not the issue is numbered separately or continuously through a single volume. This practice reduces confusion and helps the reader find the appropriate issue more efficiently.

JOURNAL ARTICLE WITH TWO OR MORE AUTHORS

The first, or senior, author is listed first, followed by the coauthor(s) in the order in which they appear on the title page:

NAME-YEAR SYSTEM	Vaughan JL, King KA, Cottrell RR. 2004. Collegiate athletic trainers' confidence in helping female athletes with eating disorders. J Athl Train. 39(1):71-76.
NUMBER SYSTEMS	24. Oyama Y, Craig RM, Traynor AE, Quigley K, Statkute L, Halverson A, Brush M, Verda L, Kowalska B, Krosnjar N, and others. Autologous hematopoietic stem cell transplantation in patients with refractory Crohn's disease. Gastroenterology. 2005;128(3):786-789.

In CSE style, the names of *all* authors of a paper (up to a maximum of 10) are given in the References section. If there are more than 10 authors, list the first 10 followed by "et al." or "and others." Note that the single-word journal title (*Gastroenterology*) is not abbreviated. Proper nouns (e.g., Crohn's) are capitalized, along with the first word of the title of the paper.

BOOK

The rules for authors and titles of books are similar to those for journal articles. Include the edition if there is more than one edition of a book. Add the designation "editor(s)" if applicable. The place of publication precedes the publisher, separated by a colon. In the name-year system, the publisher is followed by a period. In the number systems, the publisher is followed by a semicolon. In all CSE systems, the total number of pages follows at the end.

NAME-YEAR
SYSTEM
Danforth DN, editor. 1982. Obstetrics and gynecology. 4th ed. Philadelphia: Harper and Row. 1316 p.

NUMBER
SYSTEMS
6. Sokal RR, Rohlf FJ. Biometry: the principles and practice of statistics in biological research. 3rd ed. San Francisco: WH Freeman; 1994. 880 p.

ARTICLE OR CHAPTER IN AN EDITED VOLUME

NAME-YEAR
SYSTEM
Petter JJ. 1965. The lemurs of Madagascar. In: DeVore I, editor. Primate behavior: field studies of monkeys and apes. New York: Holt, Rinehart and Winston. p. 292-319.

NUMBER
SYSTEMS
9. Halliday TR. Sexual selection and mate choice. In: Krebs JR, Davies NB, editors. Behavioural ecology: an evolutionary approach. Oxford: Blackwell Scientific; 1978. p. 180-213.

In the examples above, notice that the reference listing begins with the name of the author of the particular article or chapter, not with the editor(s) of the whole collection. Be sure to include the page numbers of the section you have cited, using a period after the abbreviation for "page": p. Write out the publisher's name in full. Also, note the British spelling of the word "behavioural" (as opposed to the American "behavioral") in the second reference. Always retain the original spelling used by any source you use.

WORK WITH ORGANIZATION AS AUTHOR OR WITH UNKNOWN AUTHOR

NAME-YEAR
SYSTEM

```
[ICPO] International Committee for
Preservation of Odonata. 2000. Guide-
lines for evaluating dragonfly habi-
tats as conservation areas. Boston:
Entomological Publishers. 12 p.
```

NUMBER
SYSTEMS

```
24. Growth records for organ pipe
cactus in the Ajo area. J Desert
Nat Hist Soc. 2003;6(1):2-3.
```

If an organization is the author, place its name in the author position. Notice, above, that if an abbreviation is used for the in-text citation of the organization, it is placed in brackets and precedes the organization's full name. If the author of a work is unknown, begin the entry with the title.

FORTHCOMING WORK

The format for articles or books that are soon to be published is similar to that for published works. Include the expected date of publication, if known.

NAME-YEAR
SYSTEM

```
China-lai V. 2006. Mosses of Long
Island. Shoreham (NY): Shoreham
University Press. Forthcoming.
```

NUMBER
SYSTEMS

```
5. Johnson MR. Dietary studies of
female office workers. J Nutr Ed.
Forthcoming.
```

THESIS OR DISSERTATION

NAME-YEAR
SYSTEM

```
Grapard U. 2001. The floral biology
of Linaria vulgaris (Scrophulari-
aceae) [dissertation]. Westbury
(NY): Phipps College. 101 p.
```

NUMBER
SYSTEMS

```
3. Arnold KM. Feeding preferences
in the spiny lizard (Sceloporus jar-
rovi) [MSc thesis]. Tucson (AZ):
Southwestern University; 2004. 77 p.
```

If you have additional information about the availability of a dissertation (for example, through University Microfilms in Ann Arbor, Michigan), this information can be added after the number of pages.

CONFERENCE PAPERS

In the name-year system, give the name of the author, year of publication, paper title, name of the conference, dates and place of the conference, place of publication, publisher of the proceedings, and the page numbers for the paper:

NAME-YEAR   Rith J. 2001. Plant succession on
SYSTEM   abandoned railways in rural New York
                   State. In: King T, editor. North-
                   eastern Ecology Society 6th Sympo-
                   sium; 2001 May 2-5; Syracuse, NY.
                   Boston: Northeast Press. p. 34-41.

The CSE format for the same reference using either of the number systems is similar, except that the publication date is given after the name of the publisher:

NUMBER   16. Rith J. Plant succession . . .
SYSTEMS   Boston: Northeast Press; 2001. p.
                   34-41.

COURSE HANDOUTS OR LABORATORY MANUALS

List the author's (instructor's) name, if available, along with the year, title of the handout or lab manual, your school's location and name, and the number of pages. Here are two possible entries for course materials using CSE formats. Notice that the publication information includes both the city and the state (abbreviated in parentheses), since in each case the cities are unlikely to be familiar to all readers.

NAME-YEAR   Lynes K. 2005. Biology 141: intro-
SYSTEM   duction to plant sampling methods.
                   Bluffton Falls (NC): Payne College.
                   11 p.

NUMBER   7. Ecology 204 laboratory manual.
SYSTEMS   Laurel (MD): Gorton University;
                   2005. 112 p.

Your instructor may have his or her own preferences about the format you should use here.

CSE Style: Electronic Sources

Documentation of Internet sources has become increasingly complicated. CSE style for listing electronic sources is based on guidelines from the National Library of Medicine (NLM), which has published a lengthy supplement (2001) on Internet citations (http://www.nlm.nih.gov/pubs/formats/internet.pdf). Following are examples of several types of electronic sources. Remember that if you are using a number format, rather than the name-year format, each entry would be preceded by the number you assigned to the source.

WEB SITE OR OTHER STAND-ALONE WEB SOURCE

To list a professional or personal homepage (first page of a Web site), a portion of a Web site, or similar Web documents, you need to provide many of the same elements as for a print reference. However, Internet sources also call for some additional information, as illustrated by the following examples:

```
Worldwide Dragonfly Association [homepage on the
Internet]. Hamilton (NY): International Network
of Odonatological Information (INOI); 1998, c2003
[updated 2005 Feb 2; cited 2005 Apr 8]. Available
from: http://powell.colgate.edu/wda/dragonfly
.htm
```

```
Information about influenza pandemics [Internet].
Atlanta (GA): Centers for Disease Control and
Prevention; [updated 2005 Mar 8; cited 2005 11
Apr]. [about 2 p.]. Available from: http://www
.cdc.gov/flu/avian/gen-info/pandemics.htm
```

```
Sami Karjalainen: the dragonflies of Finland
[homepage on the Internet]. Finland: Sami Kar-
jalainen; c2005 [updated 2005 Mar 6; cited 2005
Apr 8]. Available from: http://www.korento.net/
dragonflies.html
```

For such entries, give the author(s) and the title (the most prominent words on the screen), followed by "Internet" (or "homepage on the Internet") in brackets. Next provide the place of publication or origin, followed by the publisher (organization or person responsible for issuing or sponsoring the site). Notice that an organization is often the author of a Web site; sometimes, too, the organization's name serves as the title of the source.

Homepages, reports, fact sheets, and so on, appearing as part of a Web site citation may have as many as four different dates, as in the first example, above. These are the publication date (when the material was placed on the Internet), the copyright date (preceded by "c"), the latest date the site was revised or updated, and the date on which the site was viewed ("cited") by the writer. The various dates are important because Web sites tend not to be permanent documents: their addresses and even their content are subject to sudden change. In practice, not all of these dates may be posted on the Web site; however, the date you accessed the source should always be given in your entry, and usually at least one other date about publication, copyright, or revision will be available.

In the second example, which is a short report from the CDC, the approximate extent or length of the material is given, in this case as page numbers. Such information can be expressed in other ways (for example, "about 2 screens" or "103 KB") since electronic resources often lack traditional page numbers.

The last element of all of these bibliographic citations is the URL, or Web address. Make sure you reproduce this information *exactly* as it appears, so that readers have no difficulty accessing the Web site. Do not put a period after the URL. Note also the use of brackets, semicolons, and other punctuation in the above examples. As with print sources, you must be meticulous in adhering to specific and conventional uses of punctuation in your references.

ELECTRONIC VERSION OF JOURNAL ARTICLE ALSO AVAILABLE IN PRINT

```
St. John RK, King A, de Jong D, Bodie-Collins M,
Squires SG, Tarn TWS. Border screening for SARS.
Emerg Infect Dis [serial on the Internet]. 2005
Jan [cited 2005 Apr 8];11(1):[about 10 screens].
Available from: http://www.cdc.gov/ncidod/EID/
vol11no01/04-0835.htm
```

The authors' names are listed first, followed by the title of the article and then the title of the journal. As with Web sites, brackets are used to enclose information about the electronic nature of the source. The publication date comes next, followed by the date the writer retrieved ("cited") the article (in brackets), then the volume and issue number (in parentheses). The length of the article is given in brackets (here in screens instead of traditional page numbers), followed by the URL.

ARTICLE FROM AN INTERNET-ONLY JOURNAL

```
Nuorti P, Kotilainen P, Lappalainen M. Travel
associated probable case of SARS, Finland, with
commentary from Health Canada. Eurosurveillance
Wkly [serial on the Internet]. 2003 May [cited
2005 Apr 12];7(22):[5 p]. Available from: http://
www.eurosurveillance.org/ew/2003/030529.asp
```

Notice the general similarities between the above entry and the preceding one.

OTHER INTERNET SOURCES

See the 7th edition of the CSE manual for instructions on how to list other types of electronic sources.

■ APA Style

Following are some examples of entries following guidelines in the 5th edition of the APA manual for listing sources in a References section at the end of a paper. See the manual for information on citing government documents, media sources, newspaper articles, and other sources not covered here.

JOURNAL ARTICLE WITH SINGLE AUTHOR

```
Johnson, A. J. (2005). Risk-taking behaviors by
    third-grade children in five American
    schools. New York Journal of Psychology,
    16, 23-33.
```

Note that the entry for a journal article begins with the author's last name, followed by his or her initials with periods and spaces. The date appears next, in parentheses, followed by a period and the title of the article. Only the first words of the title and subtitle (along with proper nouns and proper adjectives) should be capitalized. The journal title appears next, in italics, with the major words capitalized. APA style calls for journal titles to be written out in full, not abbreviated. The volume number appears in italics after the journal title, followed by a comma, space, and then the inclusive page numbers of the article. For journals that number each issue separately (not continuously through a single volume), give the *issue number* after the volume number, in parentheses:

```
. . . Journal of the Social Sciences, 12(2), 9-13.
```

Note that the volume number is italicized, but the issue number and parentheses are not. Also notice the use of periods and commas in specific places in this (and subsequent) entries. Like CSE format, APA style requires meticulous attention to both the proper order of information and the correct punctuation.

JOURNAL ARTICLE WITH TWO OR MORE AUTHORS

The first, or senior, author is listed first, followed by the coauthor(s) in the order in which they appear on the title page. An ampersand (&) is used to indicate "and" and comes before the last author listed:

```
Eurelle, N., & Doyle, A. (2004). Factors affect-
    ing music appreciation among elementary
    school children. Psychological Studies, 71,
    41-53.
Victor, J. C., Hoopes, E., Johnson, E. C., &
    Averill, L. (2000). Coping strategies for
    new teachers. Southeastern Journal of Edu-
    cation, 12, 6-13.
```

If there are more than six authors, list the names of the first six, followed by "et al."

```
Hayes, E., Swarthout, R. E., Bryant, H., Smith,
    R. J., LaFave, C., Novak, J., et al. (2003).
    Experimental studies of motivation in chil-
    dren of divorce. Journal of Psychology and
    Education, 31, 22-30.
```

BOOK

Give the name of the author followed by the publication date, title (italicized), place of publication, and publisher. As with titles of journal articles, capitalize only the first word of the book title and subtitle, along with any proper nouns and proper adjectives. Add information about the specific edition after the title. Notice the use of a colon to separate publisher and place.

```
Arnold, J. R. (2002). Dangerous visions: Televi-
    sion and violence in American society (2nd
    ed.). New York: Greenwood Press.
```

For an edited book, add "(Ed.)." after the editor's name, or "(Eds.)." in the case of more than one editor.

```
Maxwell, C. & Maia, D. (Eds.). (1998) Science
    and scientists: Myths and realities.
    Boston: Watertown Press.
```

ARTICLE OR CHAPTER IN AN EDITED BOOK

Give the author's last name and initials, the date the edited volume was published, and the title of the particular selection, with only the first word capitalized. Do not italicize this title. Then give the editor(s), as shown below, preceded by the word "In." Note that the initials of the editors are listed before their last names. The title of the edited volume is italicized, and only the first words of the title and subtitle are capitalized. Indicate the inclusive page numbers of the article or chapter in parentheses using the abbreviation "pp." and then give the place of publication, followed by a colon and the publisher.

> Jarvis, J. (1998). Differences in creative play behavior between four-year-old girls and boys. In S. Wright & K. Gunther (Eds.), *Child's play* (pp. 312-341). Boston: New England Press.

WORK WITH ANONYMOUS AUTHOR OR ORGANIZATION AS AUTHOR

If the author is designated as anonymous, use this word in place of the author's name. Similarly, if the author is an organization, write its name in the author position. If the same organization is also the publisher, write "Author" after the place of publication.

> Anonymous. (1971). Challenges of teaching science to elementary school children. *Topics in Education, 13*, 3-7.
> Long Island Psychological Association. (2004). *Recent trends in the hiring of psychologists.* New York: Author.

FORTHCOMING WORK

For journal articles that have been accepted but not yet published, type "in press" in parentheses instead of the year, even if the expected date is known. After the title, give just the name of the journal.

> Mahoney, C. (in press). Effects of pet ownership on play behavior in children. *Psychological Reports.*

UNPUBLISHED THESIS OR DISSERTATION

Provide the author's name, the date, and the title (italicized), followed by information about the type of thesis and the college or university:

> Stand, B. (1999). *Changing attitudes of college students towards grades and grading.* Unpublished master's thesis, Clinton University, Otselic, NY.

CONFERENCE PAPERS

Give the author of the paper, the date, and the title, followed by details about the conference and publication information.

Johnson, C. (2004). New perspectives on teaching psychology to undergraduates. In D. Fitch (Ed.), *Proceedings of the Northeast Psychology Teachers Association Conference.* St. James, NY, June 1-4, 2004. New York: Collegiate Press.

If you need to reference an unpublished paper you (or someone else) presented at a conference, use the following format. Notice that the month of the meeting is added after the year, and the title of the presentation is italicized.

Campbell, M. (2005, June). *New methods for studying toy preferences of preschool children.* Paper presented at the meeting of the Southeast Psychologists Association, Hilton Head, SC.

COURSE HANDOUTS OR LABORATORY MANUALS

Following are two possible ways to reference course materials using APA style, depending on whether or not the author is listed. Your instructor may also have preferences about how to list such materials.

Giles, C. (2005). *Statistics for Biology 252 students.* Hamilton, NY: Colgate University.

Psychology 101 Laboratory Manual. (2005). Woodbury, VA: Hartford College.

APA Style: Electronic Sources

APA style for listing electronic sources generally requires the following information: author(s), publication date or date of last update, title, retrieval date (date on which you viewed the source), publication information, and the URL. If an organization is the author, list it in the author location. If no author is identifiable, begin the entry with the title. If no date is listed, use "n.d." in the date location, in parentheses. Do not put a period after the URL.

Following are some representative examples of electronic sources using APA style.

WEB SITE OR OTHER STAND-ALONE WEB SOURCE

RxList: The Internet Drug Index. (1999). *The top*
   200 prescriptions for 1999 by number of
   U.S. prescriptions dispensed. Retrieved
   April 8, 2005, from http://www.rxlist.com/
   99top.htm

If the source is a report, fact sheet, or other document within a large government or university Web site, give the name of the organization or university before the URL for the document:

National Center for Infectious Diseases. (2005,
   January 25). *Outbreak: Polio, Ethiopia.* Retrieved April 13, 2005, from the Centers
   for Disease Control and Prevention Web
   site: http://www.cdc.gov/travel/other/
   polio_wafrica.htm

ELECTRONIC VERSION OF JOURNAL ARTICLE ALSO AVAILABLE IN PRINT

Add "Electronic version" in brackets after the title, since this form of the article may contain additional data, may have a different format, or in other ways may not be identical to the print version. Along with other standard bibliographic information, include the retrieval date and the URL:

Manns, J. R., Clark, R. E., & Squire, L. R.
   (2000). Parallel acquisition of awareness
   and trace eyeblink classical conditioning
   [Electronic version]. *Learning & Memory,*
   7(5), 267-272. Retrieved April 12, 2005,
   from http://www.learnmem.org/cgi/content/
   full/7/5/267#FN5

ARTICLE FROM AN INTERNET-ONLY JOURNAL

Include the publication date (here, the date of posting on the Internet), along with the date the document was retrieved and the URL. Notice that there are no page numbers in this type of entry.

Cardemil, E. V., Reivich, K. J., & Seligman,
   M. E. P. (2002, May 8). The prevention of
   depressive symptoms in low-income minority
   middle school students. *Prevention & Treat-*

ment, 5, Article 8. Retrieved April 6,
2005, from http://journals.apa.org/
prevention/volume5/pre0050008a.html

ARTICLE OR ABSTRACT RETRIEVED FROM A DATABASE

Pepperberg, I. M., & Kozak, F. A. (1986). Object
permanence in the African grey parrot
(*Psittacus erithacus*). *Animal Learning and
Behavior, 14*(3), 322-330. Retrieved March
6, 2005, from PsycINFO database.

If you cite an electronic version of an *abstract* retrieved from a database, then specify this information in the entry:

Shipley, C. R., & Fazio, A. F. (1973). Pilot
study of a treatment for psychological de-
pression. *Journal of Abnormal Psychology,
82*(2), 372-376. Abstract retrieved April
10, 2005, from PsycINFO database.

OTHER INTERNET SOURCES

See the APA manual for detailed instructions on how to reference many other types of material available through the Internet.

■ Arrange references either alphabetically or numerically, depending on your documentation system.

CSE style

If you are using the CSE *name-year system,* remember that your references should be listed *alphabetically* according to the senior author's last name. (See the sample research paper in Chapter 4 for an example of this method.) If a particular author has published more than one work, list these works in chronological order beginning with the earliest. In the case of two or more papers published by the same author in the same year, distinguish between these both in your reference list and your text using letters (2000a, 2000b, etc.), and use the *exact* publication date to order them. For example, a paper published by R.E. Swarthout in March would be listed before a paper she published in November of the same year. If you do not know the exact publication date, then order papers published by the same author in the same year alphabetically by title.

Works by two or more different authors with the same last name are arranged alphabetically according to the authors' first initials. For example, an article by E.J. Arnold would come before one by K.M. Arnold.

If you are using the *citation-sequence system*, number and list the references in the order in which you *cite* them in the paper. On the other hand, if you are using the *citation-name* system, list the references *alphabetically*, then number them, and use these numbers as the citation numbers in the text.

In CSE style, the reference list is titled "References" or "Cited References." (In journals that do not use CSE style, the reference list is sometimes called "Literature Cited.") Entries are often typed with all lines of each entry flush with the left margin. However, for academic assignments, it is often acceptable to begin the first line of each entry at the left margin, and then indent five spaces for successive lines (see sample paper, Chapter 4). Check with your instructor about his or her preferences regarding manuscript format. Individual journals also vary with respect to this feature of the reference section.

APA Style

If you are using APA documentation style, list all cited sources alphabetically at the end of your paper according to the last name of the senior (first) author. This reference list should be titled "References." Double-space all entries; type the first line of each entry flush with the left margin and indent subsequent lines of that entry five spaces. If there are two or more works by the same author, put them in chronological order. If the author is an organization, list the source alphabetically by the name of the organization. If the author is unknown, alphabetize the source according to the first important word of the title of the work (ignoring the words "the," "an," and "a"). If the work is anonymous, alphabetize it by the word "Anonymous."

■ Proofread your reference list for accuracy, completeness, and consistency.

Whichever documentation format you use, every entry must conform as closely as possible to the guidelines. Check that every source you cite is actually *listed* in the Literature Cited or References section. Check that no source appears in the Literature Cited section that is not *cited* in your text. You must also focus on minutiae, scrutinizing with meticulous care every punctuation mark, every space, every number, every capital letter. In the case of electronic sources, the URL must be reproduced exactly and other information must be as thorough as possible. The entire task may seem tedious indeed. Just remember that all this time and effort is indispensable to the production of a well-received paper.

CHAPTER 7

Drafting and Revising

Expect to write several drafts of your paper before you are satisfied with the final product. Good writing is generally the product of careful *rewriting,* or *revising,* in which you evaluate your early attempts at organizing and expressing your ideas. In the process you end up scrutinizing the ideas themselves, as well as your own mastery of the subject. You may find that until you can express a concept clearly enough that others can understand it, you have not fully understood it yourself. Working through successive drafts also teases out *new* ideas, perhaps connections between disparate sets of data or new insights about the significance of your research. As you grope for more appropriate wording and a clearer structure, you give form and substance to thoughts still lurking in your subconscious. Gradually and sometimes painfully, you discover what it is you really want to say.

THE FIRST DRAFT

Many people assign too much importance to a first draft. Think of it as just a rough version of the paper, an exercise in organizing your thoughts and getting down the main ideas. Focus on content, not on prose style or mechanics. Because much of what you write now may not end up in the final version, it is unproductive to tinker with commas, spelling, or sentence construction. First get down the basic information; in later drafts you can look more closely at your writing style and such matters as grammar and punctuation.

How do you shift from doing research or taking notes to composing the first draft of the paper? Experienced writers use a variety of methods. The following ones may help you get started.

■ Devise a working title.

The title should reflect your most important findings and your purpose in writing the paper. What is the main point you wish to make? What new information or perspective does your study contribute? Deciding on a preliminary title early in the writing process helps you focus your thoughts and start drafting the paper. If necessary, you can revise the title later as your ideas take shape.

■ Make an outline or a rough plan.

All writers need some sense of how the various points they want to make are logically connected to one another. However, people vary tremendously with respect to how organized their thoughts must be before they start to write. Some writers find that a detailed, point-by-point outline is absolutely essential before they start the first draft. Others prefer to work out a more general framework that leaves them considerable flexibility. Some writers find any kind of outline confining. Instead, they plunge directly from their notes into the first draft, doing most of the organizing in their heads as they go along.

Systematic planning may be more appropriate for some sections of the paper than for others. For example, the Materials and Methods section of a research paper is fairly straightforward and contains many details that must be organized in a logical manner. Once you plot it out carefully, it may practically write itself. The same may be true of the body of many review papers. By contrast, the complexities of a Discussion section may be difficult to pin down and categorize. In fact, you may not be sure of what you want to say until you start putting words to paper and have worked through one or two drafts.

It pays to experiment with different forms of planning. If you don't like outlines, try jotting down a rough list of the major topics or issues you need to discuss. You may find it easier to visualize the relationships between ideas if you make a flowchart with lines, arrows, or brackets linking groups of words or phrases. If you do rely on detailed outlines, remember that they should not be inflexible. They can be stretched or restructured to conform with your writing as it develops. An outline can be a valuable tool to get you started, but it should never be an end in itself. It should never keep you very long from that most important act of plunging into the paper.

■ Start the easiest writing first.

Don't feel you have to begin at the beginning. If you are writing a research paper, for example, it may not be productive to start with the Abstract or even with the Introduction. Begin with whatever part seems easiest and most straightforward. For many people this is the Materials and Methods section, which often can be drafted before the research is completed. The Results section also may be relatively easy to put together. Generally, the Abstract is best written last, after all other major sections of the paper have been completed. The point is that once you start drafting *any* section of the paper, you have broken the barrier between you and the material, and the writing will gain momentum.

■ Remember your audience.

Readers of any piece of writing have expectations. If you understand these expectations, you will have a clearer sense of your own purpose as a writer. You'll also be in a better position to make informed choices about the appropriate content, format, tone, and diction of your paper. Many beginning writers struggle with their first drafts because they are unsure of their intended audience. It is difficult to write in a vacuum.

For academic assignments, of course, an important reader is your instructor. Nevertheless, think of your audience more broadly as including your classmates—in other words, people with biological backgrounds similar to your own. Your peers already have some common knowledge of your subject, so you do not have to explain every single term to them. Like other scientific readers, they expect that you will meet your stated objectives, communicate your ideas clearly, develop your main points in some depth, and follow scientific conventions in organization and documentation.

For some assignments, your instructor may ask that you compose the paper as if you were writing for scholarly journals in your subject area. If your paper is aimed at a particular journal, you'll need to study its instructions to authors (often available on the journal's Web site) and follow these meticulously.

A good way to understand the general expectations of scientific audiences is simply to *read* as many scientific papers as you can. Read not just for content, but also for diction, tone, and format. If you are assigned a review paper, for example, then study examples of published reviews, even ones not directly related to your topic. Writers learn and grow not only by writing and rewriting, but also by analyzing how other writers deal with similar problems.

Finally, for works aimed at nonscientific or heterogeneous audiences, you'll need to make different decisions about style, tone, content, and organization. For example, a piece for your campus newspaper or an article in the style of a popular magazine would warrant a more informal tone. You might have a catchy headline or title. You would probably give more background information about your topic and avoid specialized terminology. Your writing might incorporate personal experience, anecdotes, direct quotes, or even dialogue—all depending on the specific venue for which you are writing. In these situations, too, it helps to study similar works by other writers. See Chapter 10 for more advice about writing for other types of audiences.

■ Talk to others.

Brainstorming with other people about the project may help get you started by giving you an audience on which to test out your ideas. Professional scientists rely heavily on such feedback from their colleagues, both in the more formal settings provided by conferences and on a day-to-day basis during lunch or coffee breaks. Electronic mail and Internet discussion groups also provide a forum for informal conversation. Intellectual exchange with others helps scientists keep their work on track and instills it with fresh insights. If you read the Acknowledgments portion of several scientific papers, you will see how important such discussions are to research and writing processes.

PRACTICAL SUGGESTIONS FOR REVISING

■ Break down the task into several successive steps.

First, consider the paper as a whole, checking overall content, organization, coherence, and consistency. Try to resist the impulse to make small-scale changes—for example, changes in punctuation—until you are satisfied with the main substance and structure of your paper. This is particularly important when you are composing on the computer. Because word processing makes it so simple to alter the surface features of your text, you may be tempted to make even your first draft "perfect," with every comma in place and every typographical error corrected. However, doing so will distract you from the more important job of articulating and organizing your main ideas.

Once you are satisfied with the paper's content and general organization, work on polishing your prose style, examining individual paragraphs and sentences closely for clarity, accuracy, and conciseness. Look for wordiness, awkward sentence constructions, improper word usage, faulty subject-verb agreement, excessive use of the passive voice, and other problems that may obscure your meaning and impede the flow of your writing. Finally, proofread the text for punctuation, spelling, and typographical errors and inspect the final form of the manuscript.

In practice, of course, making revisions is not quite so orderly and mechanical a process. Also, "writing" and "revising" a paper are not totally separate activities: writers typically make changes as they are composing, not just after they have written a draft all the way through. Nevertheless, your writing may be more productive if you set some priorities for yourself, focusing on big changes first, small changes later.

Finally, allow plenty of time for revising and proofreading your paper. Set each draft aside for several hours or several days; when you return to your writing you'll view it more objectively and work more productively. Try reading a draft out loud; you will be amazed at how useful this technique is for spotting awkwardly worded or ungrammatical constructions. Also, give your writing to others to read; see the section on peer reviews (pp. 172–174).

■ Find specific strategies that work for you.

Some writers work best if they do their early composing in longhand and save the later stages of revising and proofreading for the word processor. Others need the familiar hum of the computer right from the start and prefer to see their writing appear neatly on the screen in front of them. If you do write out early drafts in longhand, the "cut-and-paste" method saves time and effort when you need to move chunks of text from one place to another. Use scissors and tape to patch the new version together, rather than write everything over again. Of course, word processing allows writers to move words, sentences, and sections of text and to make sweeping changes in the general structure of the manuscript.

Do not discard your early drafts; keep everything at least until you have submitted the final version of the paper. Sometimes parts of a first or second draft end up sounding better than later drafts. Put both the page number and the number of the draft on each piece of writing in case you mix up parts of different drafts.

Do not confine all your revising to the computer screen. The computer shows you only a small part of the manuscript at one time, but some problems may be more noticeable when you see several pages of the paper spread out in front of you. For this reason, it makes sense to print your work frequently and make additional changes by hand directly on the manuscript. Keep all the printouts even after you have incorporated the changes in your file. You may need to refer to an early draft before you are finished with the paper.

During the later stages of editing and proofreading, remember to use word processing features that will save you time. For example, "search and replace" features of word processing programs are invaluable for locating specific words or phrases in the text and allowing you to correct misspellings or other errors. Suppose you realize that throughout the paper you have typed the genus *Gomphus* incorrectly as *Gomphis*. On command the computer can locate every place where *Gomphis* appears and automatically make the necessary correction.

Spell-checkers will locate misspelled words or typos such as *reserch* instead of *research,* or *studyy* instead of *study.* However, they cannot detect typographical errors that are legitimate words, such as *samples* instead of *sample.* They also can't distinguish between the various spellings of words such as *two, to,* and *too* or *principle* and *principal.* Finally, although spell-checkers have sizeable dictionaries, they may not include many of the words you are using, especially specialized terms and proper nouns. (However, many spell-checkers allow you to enter such words into their dictionaries.) Therefore, do not rely on spell-checkers to catch all errors and be sure to carefully proofread your final draft. (For more on proofreading, see p. 205.)

Finally, but most importantly, make sure you have two copies of your file—for example, one copy on the hard disk and a backup copy on a CD, Zip disk, or external hard disk. Remember to update the backup file each time you make revisions.

■ Solicit comments from others.

Professional biologists rely heavily on advice from their colleagues as they prepare manuscripts. Once papers are submitted to journals they are subjected to further scrutiny by editors and outside reviewers. This review process, as formidable as it may sound, gives authors invaluable feedback about their work and helps to ensure high standards for research and writing.

Ask one or more of your friends or classmates for feedback about your paper. Pick people who will give honest and thoughtful opinions and won't just say what they think you want to hear. Also, try to choose readers with some knowledge of science, preferably of your general subject, unless you are writing specifically for a lay audience. Here are some tips to make this peer review process work smoothly and effectively.

Responsibilities of the author

- Give your readers some *context* for your paper by explaining the assignment, your instructor's expectations, and your own objectives.

- Work with a *rough draft*—don't ask readers simply to proofread your writing for "errors" half an hour before the paper is due. Later you may wish to ask someone to double check for typos and other small-scale problems, but the best use of peer reviews is to help you during the early stages of the writing process, when you are still struggling with style, tone, content, and organization.

- *Orient and focus* your readers by listing your major concerns about your writing at this point. Doing so will direct their attention to key issues or places in the paper and help them give focused, specific feedback. Of course, they should also feel free to comment on other points you haven't raised.

- *Accept suggestions and criticisms graciously.* Don't get defensive. Your readers are simply trying to help. You may not agree with every piece of advice, and in fact you may receive conflicting advice. As the writer, you must take the ultimate responsibility for deciding which suggestions to follow.

Responsibilities of the peer reviewer

- *Understand the writer's chief concerns*, as well as the context and purposes of the assignment.

- *Be focused and specific in your comments.* Avoid global statements such as "This is confusing" or "I think this is really good." What points are confusing? Where did you start to get lost? In what specific ways is the paper good? Is it clear and well organized? Does the writer write with authority? Are tone and diction appropriate for the assignment? Write specific comments directly on the draft at places where they are most appropriate.

- Unless directed otherwise, *focus on global issues* rather than on misplaced commas, typos, etc. The writer is still composing big sections of text, not editing for grammar and punctuation in sentences that might be deleted anyway. However, if you notice recurrent problems—for example, sentence fragments or wordiness—that impede readability or are likely to persist in later drafts, then point these out.

- Resist the impulse to reword or otherwise "fix" the draft. Your role is to *identify possible problem areas*, not to rewrite or cowrite the paper.

- *Look for both strengths and weaknesses.* Most pieces of writing have both. The writer needs praise for successful parts of the paper as well as help identifying unsuccessful sections.

- *Be tactful* in your comments. Watch your tone; writers can have fragile egos. Give criticism in a supportive and courteous way.

- Supplement your specific in-text comments with a *brief summary of your responses* at the end of the draft. If you are one of several readers, add your name so that the writer can check with you later if he or she is confused about anything you wrote.

- Finally, *talk with the writer* informally about the paper. Listen to his or her concerns and share your own responses orally.

Following is a sample peer-reviewed draft of the introduction to a student review paper, which follows the CSE citation-sequence style of documentation.

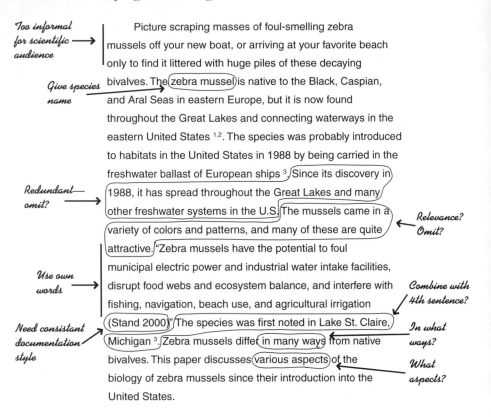

Too informal for scientific audience →

Give species name →

Redundant— omit? →

Use own words →

Need consistant documentation style →

Picture scraping masses of foul-smelling zebra mussels off your new boat, or arriving at your favorite beach only to find it littered with huge piles of these decaying bivalves. The zebra mussel is native to the Black, Caspian, and Aral Seas in eastern Europe, but it is now found throughout the Great Lakes and connecting waterways in the eastern United States [1,2]. The species was probably introduced to habitats in the United States in 1988 by being carried in the freshwater ballast of European ships [3]. Since its discovery in 1988, it has spread throughout the Great Lakes and many other freshwater systems in the U.S. The mussels came in a variety of colors and patterns, and many of these are quite attractive. "Zebra mussels have the potential to foul municipal electric power and industrial water intake facilities, disrupt food webs and ecosystem balance, and interfere with fishing, navigation, beach use, and agricultural irrigation (Stand 2000)" The species was first noted in Lake St. Claire, Michigan [3]. Zebra mussels differ in many ways from native bivalves. This paper discusses various aspects of the biology of zebra mussels since their introduction into the United States.

Relevance? Omit? ←

Combine with 4th sentence? ←

In what ways? ←

What aspects? ←

Reviewer's comments at end of paragraph:

Zebra mussels (what is the species?) sound really interesting—but I'm not sure exactly what you are going to be discussing here. Differences between zebra mussels and native bivalves? (If so, what kinds of differences?) Environmental or other impacts? Do you plan to discuss possible control measures? Perhaps you could expand statement of your aims and scope (last sentence). Have you thought of a possible title yet? I've noted places where you might omit some sentences because they are repetitive or seem (to me) unrelated to your topic. Also, for a scientific review, I think you need to have a more formal tone (I'd omit first sentence), avoid direct quotes, and use the same documentation style throughout—are you using CSE citation-sequence format? Hope these comments help—I'll look forward to seeing your next draft.

Note how the reviewer has added specific notes in the text of the draft, as well as summary comments at the end.

CHECKING CONTENT AND STRUCTURE

■ Improve logic, continuity, and balance.

In a research paper you need to keep track of your argument. What major question are you addressing? What hypotheses have you attempted to test? Do the data you present in your Results directly relate to these hypotheses and to the conclusions put forth in the Discussion? A good scientific paper is somewhat like a mystery story with the "solution" well supported by a carefully crafted body of evidence in the Results section. However, the scientific writer has also gradually prepared the reader for the major conclusions, which should come as no sudden surprise.

Although a review paper has a less formal structure, it also needs to be built around a logical train of thought. Readers need to understand your rationale for choosing the topic. They need a sense of movement as you develop your main points, and they need to see the paper end on a satisfying note through summarizing statements and conclusions.

Once you have completed a rough draft, try to visualize the paper as a unit. Is its structure clear? Do you lead readers along your line of reasoning point by point, paragraph by paragraph, making clear transitions from one topic to the next? Or do they have to grope their way along, guessing the connections? Are any topics underdeveloped or overdeveloped, throwing the paper out of balance? Do you maintain a consistent style and approach? Compare your Introduction with the rest of the paper. Have you delivered to the reader what you promised at the beginning? Are your initial statements compatible with your concluding ones?

Sometimes large-scale revisions can be accomplished relatively effortlessly by moving sentences or even whole paragraphs from one place to another. Usually, however, you need to do substantial rewriting—paring down overwritten passages, modifying highly speculative comments, developing poorly expressed ideas, adding new clarifying material, and revising the beginnings and ends of paragraphs so that you can fit them into new contexts. Such revisions can be frustrating, even painful. However, they lead to a new product: a clearer, more organized version of the paper.

■ Omit unnecessary material.

Even when you feel you have "nothing to say," you may end up with more material than you anticipated, perhaps more than you really need. As you check the manuscript for logic and balance, be alert for places where you strayed off the topic. The Materials and Methods section may include procedures that do not pertain to any of the data discussed. The Results can easily become a grab bag of miscellaneous data, only some of which are important to the main story. The Discussion may be weighed down by predictions impossible to test in the near future or by rambling comments of only peripheral importance. Check that *every* point is relevant to your main objectives. Be ruthless. Cut any extraneous material—text, table, or figure—no matter how hard you worked on it.

■ Check for completeness and consistency.

Scientific writing requires you to keep track of many details. It is surprisingly easy to create inconsistencies or to leave out pieces of needed information. This is why it's important to leave enough time to put your work aside and return to it later for a fresh look. The checklists in Chapters 3, 4, 5, and 10 may help you check the contents of tables and figures, lab reports, research papers, review papers, oral presentations, and posters, all of which have specific requirements and formats of their own.

IMPROVING PARAGRAPHS

■ Present coherent units of thought.

Paragraphs are not just chunks of text; at their best, they are logically constructed passages organized around a central idea often expressed in a *topic sentence*. A writer constructs, orders, and connects paragraphs as a means of guiding the reader from one topic to the next, along a logical train of thought.

Topic sentences often occur at the beginning of a paragraph, followed by material that develops, illustrates, or supports the main point:

> *The teeth of carnivorous and herbivorous vertebrates are specialized for different ways of life.* Those of carnivores are adapted for capturing and subduing prey and for feeding largely on meat. Dogs and cats, for example, have long, sharp canines used for piercing and molars and premolars equipped for cutting and tearing. By contrast, herbivores such as cows and horses have teeth specialized for feeding on tough plant material and breaking down the indigestible cellulose in plant cell walls. Their molars and premolars have large, ridged surfaces useful for chewing, gnawing, and grinding.

Do not distract the reader by cluttering up paragraphs with *irrelevant* information. For example, in the passage below, the second sentence does not relate to the paragraph's main point as stated in the topic sentence preceding it. We can strengthen the paragraph by deleting the second sentence altogether.

> Tanner (1981) sheds new light on the processes that may have been critical in the evolution of early hominids from chimpanzee-like ancestors. *Tanner also includes much interesting information about chimpanzee social life.* She suggests that a critical innovation might have been the extensive use of tools (initially, organic materials such as bones or sticks) by females for gathering. Such tools, she

argues, would have been employed long before weapons were being manufactured by men for hunting large game.

How long should a paragraph be? There is no set rule; paragraph length depends on the writer's topic, coverage, format, and purpose. As a working rule, aim for four to six sentences, then shorten or lengthen the passage as needed. Avoid one-sentence paragraphs; used skillfully, they give variety and emphasis, for example, in informal essays or in fiction. However, a very short paragraph may express a poorly developed idea or one that really belongs in a neighboring paragraph. Conversely, a very long paragraph can become unwieldy and confusing; usually such passages contain more than one idea and can be divided.

Keep in mind that it is not enough to group related sentences together into a paragraph. You must also *demonstrate* these relationships; otherwise, the reader cannot follow your line of thought. Look at the following passage from a lab report:

> Plants provide a constant supply of oxygen to our atmosphere. Both plants and animals depend on oxygen in the utilization of their food. During the process of photosynthesis, plants consume carbon dioxide and release oxygen.

The reader has to struggle to find the point of this paragraph. One problem is that the sentences seem to be thrown together haphazardly. Another is that the writer provides no conceptual links from one sentence to the next. Here is a revised version of the paragraph:

> During the process of photosynthesis, plants consume carbon dioxide and release oxygen. Both plants and animals depend on oxygen in the utilization of their food. Thus, plants provide a constant supply of this needed substance to our atmosphere.

In the revision the sentences are arranged more logically. Moreover, transitional elements—*thus* and *this needed substance* (referring to oxgen)—help link the last two sentences and clarify the main point of the paragraph.

Here is another paragraph to illustrate the importance of showing the reader a clear pathway of thought. Notice again the use of transitional words, along with the repetition of selected elements, to clarify the relationship between sentences:

> According to sociobiological theory, the production of individual ova is costly relative to the production of sperm. *Therefore*, ova (or females) are the limiting factor in male reproductive success, and natural selection will favor those males who can compete effectively with other males for fertilizations. We can predict that males *will vary* greatly in fitness depending on their competitive abilities. *By contrast*, females *will vary* in their abilities to convert available resources into gametes and ultimately into viable offspring.

■ **Make paragraphs work as integrated parts of the text.**

Apart from the Abstract, most paragraphs are not isolated entities. Instead, they must mesh together smoothly as structural components of each section of your paper. This means that the beginning of one paragraph must "fit" with the end of the previous paragraph—you need to bridge the gap between the two passages gracefully. If you jump abruptly from one topic to the next, the text will seem choppy and disorganized and the reader will be confused. This problem often crops up in the Results and the Materials and Methods sections when you are presenting many different kinds of information.

You can use the same techniques for linking sentences *within* paragraphs for making transitions *between* paragraphs. These include the repetition of key words or ideas and the use of transitional "markers" (*furthermore, for example, a second point, by contrast, on the one hand . . . on the other hand*, and so on) to signal to the reader that you are either developing the same idea or moving on to a different one.

■ **Vary your sentences.**

Pay attention to the structure, length, and rhythm of your sentences. People "hear" writing as they read; if your prose is unvarying and one-dimensional, you will not get your message across as effectively. The following paragraph is dominated by short, choppy sentences:

> Many doves showed "nest-calling" behavior. They assumed a position with the tail and body axis pointing slightly upwards. In this posture they flicked their wings. This behavior was seen in both sexes. It was especially common in males. I saw it performed both on and off the nest.

We can make this passage more readable (and therefore more interesting) by combining related sentences. This eliminates the distracting, choppy style and makes the paragraph more effective as a unit:

> Many doves showed "nest-calling" behavior in which they assumed a position with the tail and body axis pointing slightly upwards and flicked their wings. This behavior, especially common in males, was performed both on and off the nest.

WRITING CLEAR, ACCURATE SENTENCES

■ **Use words that say precisely what you mean.**

Do not give in to the temptation to use a word that "sounds right" unless you are absolutely sure it is appropriate. Sentences that are otherwise perfectly effective can be ruined by a single word or phrase that is wrong for the context. Here are some examples from student papers for a mycology class:

INCORRECT	Kohlmeyer (1975) sees the pit plug of red algae as being *heavily* related to the plug of Ascomycetes.
CORRECT	Kohlmeyer (1975) sees . . . *closely* related . . .
INCORRECT	Many mycologists have spent years *researching* for data suggesting a red algal ancestry with higher fungi.
CORRECT	Many mycologists have spent years *searching* . . .
INCORRECT	The evolutionary origin of the higher fungi has *harassed* scientists for many years.
CORRECT	The evolutionary origin . . . has *puzzled* scientists . . .

Buy a dictionary and consult it often. Before you use a technical term, make sure you understand its meaning. If in doubt, look it up in the glossary of an introductory text. Or check the reference section of your library for specialized dictionaries and scientific encyclopedias.

The following words are often used incorrectly in biological papers:

Affect: (as a verb) to influence or to produce an *effect*.

Effect: (as a noun) result; (as a verb) to bring about.

Nutrient concentration was the most important factor *affecting* population size.

Marking each ant on its thorax with enamel paint produced no apparent *effect* on its behavior.

We hope that further studies of these endangered species will *effect* a major change in the allocation of funds by the federal government.

Comprise: to contain or include. Do not use *comprise* when you should use *constitute* or *compose*.

INCORRECT	The vertebrate central nervous system is *comprised* of the brain and the spinal cord.
CORRECT	The vertebrate central nervous system *comprises* the brain and the spinal cord.
CORRECT	The vertebrate central nervous system is *composed* of the brain and the spinal cord.
CORRECT	The brain and the spinal cord *constitute* the vertebrate central nervous system.

Correlated: In scientific writing, use in conjunction with certain statistical tests (correlation analyses) that provide a measure of the strength of relationship between two variables. (See also p. 88.)

Interspecific: *between or among* two or more different species.

Intraspecific: *within* a single species.

Leone (1960), studying four species of sandpipers at a Minnesota lake, found marked *interspecific* differences in food preferences.

This plant shows little *intraspecific* variation in flower coloration; generally, the petals are pale yellow with a distinct orange spot at each tip.

Its/It's: *Its* is the possessive form of *it*. Do not confuse it with *it's*, a contraction of *it is* or *it has*.

Each calf recognized *its* own mother.

It's not clear whether rainfall or temperature is the more important factor.

Note, however, that contractions are generally avoided in formal biological writing.

Principal: (as an adjective) most important. Do not confuse it with *principle*, a noun, meaning a basic rule or truth.

The *principal* finding of this study was that excessive drinking by rats did not cause significant increases in blood pressure.

According to Heisenberg's Uncertainty *Principle*, we can never simultaneously determine the position and the momentum of a subatomic particle.

Random: In scientific writing, use to refer to a *particular* statistically defined pattern of heterogeneous values. (See also p. 88.)

Significant: In scientific writing, use to refer to *statistically significant* (or nonsignificant) results, after having conducted appropriate statistical tests. (See also p. 87.)

That/Which: Use *that* to introduce *restrictive* or defining elements— phrases or clauses that limit your meaning in some way.

The rats *that had been fed a high calorie diet* were all dead by the end of the month.

Here the italicized portion restricts the meaning of *rats;* we are referring to *only* those specific rats that had been fed a high calorie diet.

Use *which* to introduce *nonrestrictive* or nondefining elements—word groups that do not limit your meaning but rather add additional information. Because this information is not vital to the integrity of the sentence, you can omit it without substantially changing the original meaning. The following sentence uses *which* instead of *that;* here the writer is speaking more generally, not calling attention to a *particular* group of rats.

The rats, *which* were fed a high calorie diet, were all dead by the end of the month.

Misuse of *that* or *which* may make a sentence confusing:

Plants, *which* grow along heavily traveled pathways, show many adaptations to trampling.

Here it sounds as if *all* plants grow along heavily traveled pathways. Actually the writer is referring only to a particular group of plants. The nonrestrictive clause needs to be replaced by a restrictive one limiting the meaning of *plants:*

> Plants that grow along heavily traveled pathways show many adaptations to trampling.

Do *not* use commas with restrictive elements and *that,* as in the example above. Use commas to set off nonrestrictive elements introduced by *which*—a *pair* of them if the element appears in the middle of the sentence:

> These data, which are consistent with those of other researchers, suggest several questions about the significance of wing positions in the thermoregulation of Arctic butterflies.

Unique: one of a kind. A thing cannot be *most, very,* or *quite* unique; it is simply unique:

INCORRECT	Females of this species have a *very unique* horny projection from the dorsal part of the thorax.
CORRECT	Females of this species have a *unique* . . .

■ Avoid slang.

Slang is the informal vocabulary of a particular group of people. Some slang words eventually make their way into standard English; however, most soon become outdated and are replaced by others. Slang usually has no place in scientific writing, even if you put it in quotation marks.

SLANG	The controversy over the evolutionary origin of the Ascomycetes and Basidiomycetes dates back almost a hundred years, but it has only recently moved to the cutting edge of research.
STANDARD	Recently, scientists have become more interested in the evolutionary origin of the Ascomycetes and Basidiomycetes, a topic that has been controversial for almost a hundred years.
SLANG	Barr's (1980) statement is merely a "cop-out" because he refuses to acknowledge that there are major morphological differences between the two groups.
STANDARD	Barr (1980) fails to address this issue because he refuses to acknowledge that . . .

■ Revise misplaced modifiers.

Modifying words, phrases, or clauses should relate clearly to the words they are meant to describe. If they do not, the sentence may be confusing, even ludicrous.

FAULTY   After marking its hindwings with enamel paint, each damselfly was released within 1 m of the capture site.

REVISED   After marking its hindwings with enamel paint, I released each damselfly within 1 m of the capture site.

Unless these insects were particularly clever at marking themselves, we need to make the intended subject of this sentence (the writer) more evident. Using the active instead of the passive voice helps here. (See also pp. 189–191.)

FAULTY   After mating, the sperm are stored in a sac within the female damselfly's body.

REVISED   After mating, the female damselfly stores sperm in a sac within her body.

Who is actually mating, the sperm or the damselfly? The revised version corrects the ambiguity.

FAULTY   This behavior has only been reported in one other primate.

REVISED   This behavior has been reported in only one other primate.

Modifiers such as *only, even, almost,* and *nearly* should be placed next to the most appropriate word in your sentence, depending on your meaning. In the first sentence, the position of *only* before the verb suggests that behavior may have been observed in more than one primate but *reported* in just a single species. If you really want to say that this behavior has been observed *and* reported in just this one species, then *only* must modify *one* as in the revised version. In the following sentence, *only* does limit the meaning of the verb in accordance with the writer's meaning.

This behavior has only been observed casually, not reported formally.

■ Avoid vague use of *this, that, it,* and *which.*

Do not use these pronouns on their own to refer to whole ideas; you may lose the reader. To avoid confusion, use clear, specific wording.

VAGUE   We could not predict the number of adult males likely to visit each breeding site because male density in the surrounding forest varied greatly from day to day. This is typical of most field studies on this species.

What is "typical of most field studies"? Varying male densities, being unable to predict male numbers at the breeding site, or both? The revision eliminates this ambiguity:

> **SPECIFIC** We could not predict the number of adult males . . . from day to day. Varying male density is typical of most field studies . . .

■ Make comparisons complete.

Add words if necessary to make comparisons or contrasts accurate and unambiguous.

> **AMBIGUOUS** Average body length in *Libellula pulchella* is longer than *Plathemis lydia.*

Here, you want to compare the body length of one species with the *body length* of another, as the revision clearly does:

> **UNAMBIGUOUS** Average body length in *Libellula pulchella* is longer than that in *Plathemis lydia.*
>
> **AMBIGUOUS** Bullfrogs were more abundant than any amphibian at Site A.

Because bullfrogs themselves are amphibians, you need to revise the wording to avoid confusion:

> **UNAMBIGUOUS** Bullfrogs were more abundant than any other amphibian at Site A.

■ Make each verb agree with its subject.

Do not lose sight of the subject in a sentence by focusing on modifying words, such as prepositional phrases, occurring *between* the subject and the verb.

> The *size* of all territories *was* [not *were*] reduced at high population densities.
>
> The *zygote* of the Ascomycetes *develops* [not *develop*] into ascospores.

Both sentences above have singular subjects and therefore need singular verbs.

Compound subjects are subjects with two or more separate parts that share the same verb. When these parts are connected by *and*, they need to be matched with plural verb forms.

> The *color and shape* of the beak are [not *is an*] important taxonomic features [not *feature*].

When the parts of a compound subject are linked by *or* or *nor*, make the verb agree with the part that is closest to it. If one part of the subject is singular and the other plural, then put the plural part second and use a plural verb.

> Under experimental conditions, *neither* the newly hatched chick *nor* its older siblings *were tended* by the parents.

Here the parts of the compound subject are *chick* and *siblings*; these are joined by *nor*. The second part, *siblings*, is plural and needs a plural verb (*were tended*).

Do not confuse a compound subject with a singular subject that is linked to other nouns by a prepositional phrase (such as *in addition to, along with, as well as*). The following sentence has a singular subject and takes a singular verb:

> The dominant *male*, along with his subordinates, *protects* [not *protect*] the offspring when the troop is threatened by predators.

When you refer to a particular quantity of something as a single unit, treat it as singular.

> Before each experiment, *10 ml* of distilled water *was* added to each vial.

Many scientific terms have a Latin heritage and are used in both the singular and plural form, depending on the context. Make sure the verb agrees with the subject.

> The *larva* of the monarch butterfly *feeds* [not *feed*] on milkweed.

Here are the singular and plural forms of some words commonly used in biology.

SINGULAR	PLURAL
alga	algae
analysis	analyses
bacillus	bacilli
bacterium	bacteria
basis	bases
criterion	criteria
datum (*rarely used*)	data
flagellum	flagella, flagellums
focus	foci, focuses
formula	formulae, formulas
fungus	fungi, funguses
genus	genera
hypothesis	hypotheses
index	indices (for numerical expressions), indexes (in books)

inoculum	inocula
larva	larvae
locus	loci
matrix	matrices
medium	media, mediums
mycelium	mycelia
nucleus	nuclei
ovum	ova
phenomenon	phenomena, phenomenons
phylum	phyla
protozoan	protozoa, protozoans
pupa	pupae
serum	sera, serums
species	species
spectrum	spectra
stimulus	stimuli
stratum	strata
symposium	symposia, symposiums
taxon	taxa
testis	testes
villus	villi

Note that the word *species* can be either singular or plural: one species of cat, three species of toads.

■ Put related elements in parallel form.

When you link two or more words, phrases, or clauses in a sentence, put them in the same grammatical form. Such parallelism makes your writing easier to read and emphasizes the relationship between items or ideas.

FAULTY	These two species differ in color, wingspan, and where they typically occur.
PARALLEL	These two species differ in color, wingspan, and habitat.
FAULTY	Both populations of plants had high mortality rates at sites that were dry, windy, and where there were frequent disturbances.
PARALLEL	Both populations of plants had high mortality rates at sites that were dry, windy, and frequently disturbed.
FAULTY	Male bluntnose minnows promote the survival of their offspring by agitation of the water over the eggs and keeping the nest free from sediment.

PARALLEL   Male bluntnose minnows promote the survival of their offspring by agitating the water over the eggs and keeping the nest free from sediment.

In the first example, the clause, *where they typically occur*, has been replaced by a single noun, *habitat*, for parallelism with the two preceding nouns (*color, wingspan*). Similarly, in the second example, parallelism has been achieved by putting *where there were frequent disturbances* into adjective form, *disturbed*, to agree with *dry* and *windy*. In the third example, the noun *agitation* has been changed to a present participle (verb form ending in *-ing*), making two parallel phrases, "*agitating . . .*" and "*keeping. . . .*"

■ Write in a direct, straightforward manner; avoid jargon.

Scientific writing has the reputation of being dry, monotonous, and hard to understand. Consider the following passage from a published paper.

> The data of this study suggest that such a handling procedure not only effects a diminution of emotional behaviour as indexed by decreases in the duration of the immobility reaction, but also as indexed by other measures of fear (freezing and mobility in the open field). To the extent that the handled and nonhandled groups differed in the predicted direction with respect to these indices of fear, distress vocalizations in the open field were shown to be significantly less frequent in the relatively more fearful (nonhandled) group. (Ginsberg, Braud, and Taylor 1974, p. 748)

A paper may contain exciting results or brilliant insights, but still be tedious to read because of the author's unengaging prose. Whether it is a manuscript submitted for publication or a laboratory report for a biology class, an impenetrable style may keep it from getting the attention it deserves.

Frequently, the problem is the use of *jargon* instead of simpler, more straightforward writing. Broadly, jargon is the technical language of some specialized group, such as biologists. More specifically (and negatively), jargon is long-winded, confusing, and obscure language. Writers of jargon use esoteric terms unfamiliar to most of their readers. They rely heavily on long sentences, big words, a pompous tone, and ponderous constructions in the passive voice.

JARGON   One hour prior to the initiation of the experiment, each avian subject was transported by the experimenter to the observation cage. The subject was presented with various edible materials, and ingestion preferences were investigated utilizing the method developed by Wilbur (1965). When data collection was finalized, the subject was transferred back to the holding cage.

REVISED	One hour before the experiment, I put each bird in the observation cage, where feeding preferences were studied using Wilbur's (1965) method. The bird was then returned to the holding cage.

The revised passage conveys the same information as the jargonridden one, but more simply, directly, and concisely.

Do not assume that to sound like a biologist you must write dry, stilted, and boring prose or that complex ideas must be couched in equally complex, convoluted sentences. Biology instructors, along with editors and readers of biological journals, prefer clear, straightforward writing—simple but effective prose that quickly illuminates the author's results and ideas.

AVOIDING WORDINESS

First drafts are usually labored and wordy because you have been focusing on just writing down your ideas. As you revise, examine each sentence carefully. Could you say the same thing more succinctly without jeopardizing the content? Lifting just one excess word from a sentence can enliven its rhythm and intensify its meaning, making your prose carry more weight.

■ **Omit unneeded words; shorten wordy phrases.**

WORDY	On two occasions, I succeeded in observing a mating pair for the entire duration of copulation.
CONCISE	I observed two pairs for the duration of copulation.

WORDY	There now is a method, which was developed by Jones (1973), for analyzing the growth of rotifer populations.
CONCISE	Jones (1973) developed a method to analyze the growth of rotifer populations.

WORDY	The sample size was not quite sufficiently large enough.
CONCISE	The sample size was not large enough.
MORE CONCISE	The sample size was too small.

WORDY	The root cap serves to protect the cells of the root meristem as the root is growing through the soil.

CONCISE	The root cap protects the cells of the root meristem as the root grows through the soil.
WORDY	The eggs were blue in color, and they were covered with a large number of black spots.
CONCISE	The eggs were blue with many black spots.

Common modifiers such as *very, quite,* and *rather* can often be cut from sentences. If you use such words routinely, ask yourself if they are essential to your meaning.

FAULTY	The data in Table 1 are *very consistent* with Leshchva's (1966) model.
REVISED	The data in Table 1 are *consistent* with . . .
FAULTY	Males guarding eggs are *quite aggressive* toward juveniles and females.
REVISED	Males guarding eggs are *aggressive* toward . . .

In summary, biological writing is plagued by countless wordy phrases, often placed at the beginning of a sentence, and by "empty" words and phrases that add little to the author's meaning. Some of these are listed below. You will probably think of many others to add to the list.

WORDY	CONCISE
a second point is that	second, secondly
more often than not	usually
it is apparent that	apparently
at the present time	now
in previous years	previously
owing to the fact that	because
because of the fact that	because
in light of the fact that	because
it may be that	perhaps
these observations would seem to suggest	these observations suggest
one of the problems	one problem
in only a very small number of cases	occasionally, rarely
in the possible event that	if
An additional piece of evidence that helps to support this hypothesis	Further evidence supporting this hypothesis
In spite of the fact that our knowledge at this point is far from complete	Although our present knowledge is incomplete

It is also worth pointing out that	*omit it*
Before concluding, another point is that	*omit it*
It is interesting to note that	*omit it*

■ Avoid repetition.

Some sentences or paragraphs are wordy because the writer has included the same information twice.

| WORDY | In Kohmoto's study in 1977, she failed to account for temperature fluctuations (Kohmoto 1977). |
| CONCISE | Kohmoto (1977) failed to account for temperature fluctuations. |

Because the author's name and the publication date are given in the literature citation (Kohmoto 1977), you need not give the same information at the beginning of the sentence. (See also Chapter 6.)

In the next example, the second sentence can be omitted because it merely repeats part of the first sentence using different wording.

Male fathead minnows who were tending eggs spent the majority of their time rubbing the egg batches with their dorsal pads and preventing other fish from eating the eggs. They did not devote much time to other activities, but instead were chiefly occupied with behavior directed toward the eggs.

■ Use the passive voice sparingly.

In the passive voice, the subject of the sentence *receives* the action, whereas in the active voice it *does* the action.

| PASSIVE | Nearly half the seedlings *were eaten* by woodchucks. |
| ACTIVE | Woodchucks *ate* nearly half the seedlings. |

Biological writing leans heavily on the passive voice even though the active voice is more direct, concise, and effective. Why this emphasis on the passive? Compare the following two sentences:

| PASSIVE | Skin extract solution *was presented* to the fish through a plastic tube. |
| ACTIVE | *I presented* skin extract solution to the fish through a plastic tube. |

Neither sentence is incorrect; however, the sentence in the passive voice shifts the reader's attention away from the writer and more appropriately to the materials he or she has been testing, observing, collecting, or measuring. This is why the passive voice is particularly common in the Materials and Methods section of biological papers. Shunning the passive altogether may result in prose that is "I-heavy" and monotonous because of too many first-person references. Prudent use of passive constructions gives variety to the text, and in a few situations such constructions may simply be more convenient.

Thus the passive voice has legitimate uses. Unfortunately, however, many beginning writers rely heavily on the passive because they think it makes their prose more formal, more important, more "fitting" for a scientist. When used habitually, carelessly, or unintentionally, the passive voice results in a wordy and cumbersome style. Overuse of the passive is one reason that scientific writing has the reputation of being dry, pompous, and boring. Usually the passive voice can easily be converted to the active voice, making sentences shorter and more forceful without any loss of meaning.

PASSIVE	Territory size was found to vary with population density.
ACTIVE	Territory size varied with population density.
PASSIVE	From field observations, it was shown that virtually all tagged individuals remained in their original home ranges.
ACTIVE	Field observations showed that virtually all tagged individuals remained in their original home ranges.
PASSIVE	Nest destruction was caused primarily by raccoons, particularly late in the incubation period, when greater access to nests was afforded to them by lowered water levels.
ACTIVE	Raccoons caused most nest destruction, particularly late in the incubation period when lowered water levels afforded them greater access to nests.

The passage below, from the Methods section of a paper by Burger (1974, p. 524), illustrates an effective mix of active and passive voice.

> I used several marking techniques on nests, adults, and juveniles. In 1969, I marked nests with red, blue, and white plastic markers tied to cattails. . . . Markers placed on nests were subsequently covered with fresh nest material. Adults were captured with a nest trap . . . and marked with coloured plastic wing tags (Saflag). I marked pairs who were close to the blind by pulling a string attached to a cup of red dye suspended over the nest.

Notice that the author of this passage uses several first-person references (*I* used . . . ; *I* marked . . .). Today many biologists are writing "I" instead of impersonal and cumbersome language such as "this investigator" or "the author." This use of the first person makes their prose more direct and concise. It may also reflect a growing realization by biologists that there *is* an "I" in science—that scientific research is inevitably influenced by the personal background, interests, motives, and biases of each researcher.

You will still find the passive voice in most of the scientific literature you read. As you write your own papers, you may feel the need to use the passive voice. There are no firm guidelines about this in biological writing, and different journals and scientific fields vary in their use of the passive. However, the active voice generally does a better job. Use passive constructions deliberately and sparingly. For every sentence you put in the passive, ask yourself if you could express it more exactly and concisely in the active voice.

VERB TENSE

Scientific ethics have given rise to the convention of using the *past* tense when reporting your own present findings and the *present* tense when discussing the published work of others. This is because new data are not yet considered established knowledge, whereas the findings of previous studies are treated as part of an existing theoretical framework. Therefore, in a research paper you will need to use both the past and present tenses. Most of the Abstract, Materials and Methods, and Results sections will be in the past tense because you are describing your own work. Much of the Introduction and Discussion will be in the present tense because they include frequent references to published studies. Look at the following examples.

The reproductive success of yellow-bellied marmots (*Marmota flaviventris*) is strongly influenced by the availability of food and burrows (Andersen, Armitage, and Hoffmann 1976) [Introduction].

After inoculation, plants were kept in a high-humidity environment for 100 h [Materials and Methods].

Limpets occurred at all sampling sites [Results].

Diaptomus minutus was dominant in the zooplankton of Clinton Lake during both years of the study. This species is common on many other acidic lakes in the region (Lura and Lura 1985) [Discussion].

When you refer to an author directly, however, you may use the past tense:

Bauman (1959) found that this bacterium is highly sensitive to pH.

When you refer directly to a table, a figure, or a statistical test in your own paper, it is acceptable to use the present tense.

Table 3 shows that polychaetes were most abundant at depths of 10–16 m.

See the sample research paper at the end of Chapter 4 for other examples of correct verb tense.

PUNCTUATION

Used correctly, punctuation marks help make your writing clear and understandable. Used incorrectly, they may distract, annoy, or confuse the reader. Buy a writer's guide with a detailed section on punctuation, and consult it frequently; several good handbooks are listed at the end of this book. The guidelines below address some of the most common punctuation problems in biology writing.

Comma

1. Use a comma to separate introductory material from the rest of the sentence.

Although egg cases were reared using Hendrickson's (1977) method, none of the eggs hatched.

During the 1987 breeding season, bullfrogs were sexually active from early June to late July.

Nevertheless, these data suggest that mate selection in this species is based primarily on female choice.

A comma may be omitted after a very *short* introductory element that merges smoothly and unambiguously with the rest of the sentence.

Thus the data in Table 1 are consistent with those of Tables 2 and 3.

2. Use a *pair* of commas to set off insertions or elements that interrupt the flow of a sentence.

The situation is different, however, on isolated islands with lower species diversity.

This explanation, first proposed by Hess (1967), is still widely accepted.

3. Use a comma to separate all items in a series, including the last two.

Unlike mosses, ferns possess true roots, stems, and leaves.

4. Use a comma to separate independent clauses linked by a coordinating conjunction (*and, but, or, for, nor, so, yet*). (An independent clause is a word group containing a subject and a verb and able to function as a complete sentence.)

> Many studies have been made of feeding preferences in spiders, but few have been done under natural conditions.

Do *not* use a comma on its own to join two independent clauses. Such an error is called a *comma splice*.

> COMMA SPLICE   Horsetails usually grow in moist habitats, some occur along dry roadsides and railway embankments.

A comma splice also results when you join two independent clauses with a comma followed by a conjunctive adverb (such as *however, moreover, furthermore, therefore, nevertheless*).

> COMMA SPLICE   Horsetails usually grow in moist habitats, however some occur along dry roadsides and railway embankments.

There are several ways to correct a comma splice. For example, you can use a semicolon (see p. 194) with or without a conjunctive adverb:

> REVISION 1   Horsetails usually grow in moist habitats; however, some occur along dry roadsides and railway embankments.
>
> REVISION 2   Horsetails usually grow in moist habitats; some occur along dry roadsides and railway embankments.

A comma splice can also be corrected by adding a coordinating conjunction after the comma:

> REVISION 3   Horsetails usually grow in moist habitats, but some occur along dry roadsides and railway embankments.

Alternatively, you can make one of the clauses a dependent clause, and separate the two clauses by a comma. (A dependent clause lacks either a subject or a verb and cannot stand alone as a complete sentence.)

> REVISION 4   Although horsetails usually grow in moist habitats, some occur along dry roadsides and railway embankments.

Finally, you can make two sentences separated by a period:

> REVISION 5   Horsetails usually grow in moist habitats. Some occur along dry roadsides and railway embankments.

Semicolon

1. Use a semicolon to join two independent clauses *not* linked by a co-ordinating conjunction (*and, but, or, for, nor, so, yet*).

> Bacteria reproduce very rapidly; many species can divide once every 20 minutes under favorable conditions.

This statement could also have been written as two complete sentences separated by a period, but the semicolon suggests a closer relationship between the two ideas.

2. Use a semicolon to connect two independent clauses joined by a conjunctive adverb (such as *however, moreover, nevertheless, furthermore*).

> These data are consistent with those of Allen (2004); moreover, they suggest that pH may be a more important influence than previously believed.

3. Use a semicolon to clarify the meaning of a series of items containing internal commas.

> Films were made of courtship and mating behavior; aggressive interactions between males, including chasing, butting, and biting; and parental behavior by females, particularly fanning and egg retrieving.

Colon

1. Use a colon to introduce items in a series.

> Birds are distinguished by the following features: a four-chambered heart, feathers, light bones, and air sacs.

However, do *not* use a colon before a series unless the colon follows an independent clause.

> INCORRECT   Dandelions are in the same family as: daisies, chrysanthemums, sunflowers, and hawkweeds.
>
> INCORRECT   The most common insects at Site A were: mayflies, dragonflies, damselflies, and butterflies.

In both these examples, the colon should be deleted.

2. Use a colon between two independent clauses when the second clause explains or clarifies the first one.

> The Cranston Lake study site was most suitable for incubation studies: nest sites were abundant, and brooding birds were rarely disturbed by road traffic.

3. Use a colon to formally introduce a direct quotation. (See p. 126.)

Dash

The dash is used to separate material abruptly from the rest of the sentence for clarity, emphasis, or explanation. (To type a dash, use two hyphens, one after the other, with no space on either side: --. Word processing programs can then convert these hyphens to dashes.)

Insects are distinguished by three body regions—head, thorax, and abdomen—and three pairs of legs.

Many writers use the dash when a comma, indicating just a slight pause, would be less distracting.

DISTRACTING   Eggs with minor cracks and an intact inner membrane hatched normally—but severely cracked eggs failed to develop and soon rotted.

REVISED   Eggs with minor cracks and an intact inner membrane hatched normally, but severely cracked eggs failed to develop and soon rotted.

When overused, dashes make your writing seem informal and careless, suggesting that you do not know how to use other types of punctuation. Use dashes sparingly, if at all, in biological writing.

Parentheses

1. Use parentheses to insert explanatory or supplemental information into a sentence.

The nests at Leland Pond were unusually large (at least 30 cm in diameter), but only one contained any eggs.

Juvenile monkeys raised in groups of three gained weight more rapidly than those raised in isolation (Fig. 2).

Like dashes, parentheses should be used sparingly because they interrupt the flow of your writing. Commas often work just as well as parentheses and are less distracting.

2. Use parentheses with a series of items introduced by letters or numbers, especially when you are listing many items or several long items.

Dragonflies are good subjects for studies of territorial behavior for these reasons: (1) many are easily identified and widely distributed, (2) most can be observed easily under natural conditions, and (3) much is known about the anatomy and life history of many species.

Quotation Marks, Brackets, Ellipsis

See pages 124–126 for uses of these punctuation marks to present quoted material.

CHAPTER 8

Preparing the Final Draft

MECHANICS AND TECHNICALITIES

■ Use the scientific names of organisms.

The common names of plants and animals have arisen through long traditions of folklore and popular usage. Many reflect some aspect of an organism's appearance, natural history, or relevance to human life: lady's slipper, hedgehog, bedbug, horsehair worm, liverwort. However, common names can also be misleading. Ladybugs, strictly speaking, are not bugs (Hemiptera) but beetles (Coleoptera), and club mosses (Lycopsida) are not true mosses (Musci). Because common names are not universally agreed upon, a single species may be called by different names in different localities. Also, the same common name may be used for two or more taxonomically unrelated organisms.

However, each species of organism has just one scientific name. It consists of two parts: first, the word denoting the genus; and second, the specific epithet. Both typically are Latin words or Latinized forms. For example, the scientific name of red maple is *Acer rubrum; Acer* is the genus (to which many different species of maples belong). *Acer rubrum* is the particular species, red maple.

The scientist who first publishes the scientific name of a newly recognized species is regarded as its "author"; his or her last name is placed after the generic and specific names: *Pimephales promelas* Rafinesque. An author's name is often abbreviated: R. for Rafinesque, L. for Linnaeus, Fab. for Fabricius. (However, sources may vary with respect to the abbreviations used for the same author's name.) Sometimes an author's name is put in parentheses, indicating that the species has been put into a different genus from the original. In botany and microbiology, but not zoology, the name of the person responsible for this new classification is also added at the end. For example, in the case of *Kickxia spuria* (L.) Dum., the plant was originally called *Linaria spuria* by Linnaeus but was later placed in a different genus, *Kickxia*, by Dumortier (Dum.).

Thus, the scientific names of organisms are determined according to a system of universally accepted rules and conventions. Biologists rely heavily on scientific names; they use common names less frequently. (In fact, most species known to science lack common names.) It is acceptable to refer to organisms by common names as long as you have first given full scientific names along with other taxonomic information, if necessary. Here are some general guidelines for using scientific and common names:

1. The scientific names of species are always italicized (or underlined if you are writing in longhand). The genus name is capitalized; the specific epithet is not: *Plumularia setacea*. Subspecies or other names below the level of species are italicized but not capitalized: *Plectrophenax nivalis subnivalis*. Author names or their abbreviations are capitalized but not italicized: *Haematosiphon inodorus* (Duges).

2. Give the full scientific name the first time a particular species is referred to in the text. In later references to this species you may abbreviate the genus by its first letter (still capitalized and italicized). The bacterium *Spirillum volutans* thus becomes *S. volutans*. Confusion may arise, however, if you are discussing two or more organisms whose generic names begin with the same letter—for example, *Spirillum volutans* and *Streptococcus salivarius*. In that case, it is safer to spell out the generic names.

Also, spell out the generic name at the start of a sentence, or else reword the sentence.

INCORRECT	*S. vicinum* was the most common dragonfly at Site A.
CORRECT	*Sympetrum vicinum* was the most common dragonfly at Site A.
CORRECT	The most common dragonfly at Site A was *S. vicinum*.

3. Generally, the author of a species is specified only the first time the species is mentioned in the paper. Often this information is included in the title, as well.

4. Do not put an article (*the, a, an*) immediately before the scientific name of a species.

INCORRECT	The most common lichen at both sampling sites was the *Lecidea atrata*.
CORRECT	The most common lichen ... was *Lecidea atrata*.

5. Do not pluralize scientific names.

INCORRECT	Digger bees (*Centris pallidas*) were also observed at the study site.
CORRECT	Digger bees (*Centris pallida*) were ...

6. Except in keys (identification guides) or other taxonomic writings, the specific epithet of an organism must always be preceded by the generic name or its abbreviation.

INCORRECT	The most common species at Hamilton Creek were *Achnanthes minutissima* and *Meridion circulare. Circulare* was also the most common alga at Woods Creek later in the sampling period.
CORRECT	The most common species . . . were *Achnanthes minutissima* and *Meridion circulare. Meridion circulare* was also ...

The names of genera, however, may be used alone if you are referring collectively to the species in a particular genus.

Insecticides have been used in an attempt to eradicate *Anopheles* mosquitoes and thus control the spread of malaria.

Some species of *Sargassum* grow in dense mats on the surface of the ocean.

7. Taxonomic groups, or taxa, above the level of genus (family, order, class, phylum, division, and so on) are capitalized but not italicized or underlined.

The Chilopoda (centipedes) and the Diplopoda (millipedes) are two of six classes in the subphylum Mandibulata.

8. Biologists frequently drop or modify the endings of taxa to make common names for organisms—for example, chironomids from Chironomidae; lycopsids from Lycopsida; dipterans from Diptera; cephalopods from Cephalopoda. Such words, unlike the formal names for groups, are not capitalized.

9. Similarly, other common names of organisms are not usually capitalized except in accordance with specific taxonomic guidelines for certain groups. For example, the common names of American birds have been standardized by the American Ornithologists' Union and *are* capitalized: American Robin, Chipping Sparrow, Barn Swallow. Shortened forms of these proper common names are not capitalized: robin, sparrow, swallow. If a common name contains a word derived from the name for a particular person or place, then that word is generally capitalized even if the rest of the common name is not: English ivy, Queen Anne's lace.

10. Biologists use conventional abbreviations to refer to one or more undesignated species of a particular genus:

Most nematodes collected from Site 4 were affected by a fungal parasite (*Myzocytium* sp.) [one species].

Francini (1970) found that in certain butterflies (*Colias* spp.) . . . [more than one species].

■ Record time according to a 24-hour system.

Biologists do not report times as AM or PM. Instead, they use a 24-hour time system, numbering the hours of the day consecutively starting at midnight (0000 hours). For example:

7:30 AM	=	0730
12:00 noon	=	1200
9:45 PM	=	2145
11:53 PM	=	2353

Biological writing also uses a convenient shorthand to specify *photoperiod* (the number of hours of light in a 24-hour period). For example, instead of writing, "Hamsters were exposed to a photoperiod of 12 h," a writer could say, "Hamsters were reared under conditions of 12L:12D" (12 hours of light followed by 12 hours of darkness). Similarly, 14L:10D means 14 hours of light and 10 hours of darkness.

■ Use symbols and abbreviations commonly used in biology.

Following are some symbols and abbreviations commonly used in biology. Note that many are not followed by a period (unless they appear at the end of a sentence). There is some variation from one scientific journal to the next. The Council of Science Editors (CSE) style manual (see Additional Readings, pp. 253–255) contains extensive lists of recommended symbols and abbreviations for biologists.

TERM/UNIT OF MEASUREMENT	SYMBOL/ ABBREVIATION
ångström	Å
approximately	ca. *or* c. *or* ≈
calorie	cal
centimeter	cm
cubic centimeter	cm^3
cubic meter	m^3
cubic millimeter	mm^3
day	d
degree Celsius	°C
degree Fahrenheit	°F

TERM/UNIT OF MEASUREMENT	SYMBOL/ABBREVIATION
degrees of freedom	df, ν
diameter	diam
east	E
et alia (Latin: "and others")	et al.
et cetera (Latin: "and so forth")	etc.
exempli gratia (Latin: "for example")	e.g.
female	♀
figure, figures	Fig., Figs.
foot-candle	fc *or* ft-c
gram	g
greater than	>
hectare	ha
height	ht
hour	h
id est (Latin: "that is")	i.e.
joule	J
kelvin	K
kilocalorie	Kcal
kilogram	kg
kilometer	km
latitude	lat.
less than	<
liter	L *or* l
logarithm (base 10)	log
logarithm (base *e*)	ln
longitude	long.
male	♂
maximum	max
mean (sample)	$\bar{x}, \bar{X}, \bar{y},$ or $\bar{Y}$
meter, metre	m
microgram	μg
microliter	μL *or* μl
micrometer	μm
milliliter	mL *or* ml
millimeter	mm
minimum	min
minute (time)	min
molar (concentration)	mol L^{-1}
mole	mol
month	mo
nanometer	nm
north	N
not (statistically) significant	NS *or* ns
number	nr

TERM/UNIT OF MEASUREMENT	SYMBOL/ ABBREVIATION
number (sample size)	n
parts per million	ppm
percent	%
plus or minus	$\pm$
probability	P
second (time)	s
south	S
species (singular)	sp.
species (plural)	spp.
square centimeter	cm^2
square meter	m^2
square millimeter	mm^2
standard deviation	s or SD
standard error	$s_{\overline{X}}$ or SE
standard temperature and pressure	STP
versus	vs.
volt	V
volume	vol
watt	W
week	wk
weight	wt
west	W
year	y

MANUSCRIPT FORMAT

After making the final revisions in a piece of writing, you may feel that all the work is over. It is—almost. You still need to produce a neat, clean manuscript. A paper with page numbers missing, margins askew, and the type barely readable will not show off your prose to good advantage, and it may suggest that little care was taken in more substantive matters. At the very least, a carelessly prepared manuscript creates a bad first impression, and in some instances it may not be read at all. Now that computer technology has so simplified the basic, mechanical process of manuscript preparation, there simply is no excuse for a sloppily prepared document.

Biologists submitting work for publication must adhere to specific format guidelines when preparing the final manuscript. These are listed in an Instructions to Authors or similarly titled document, available online or in a special issue of each journal. Now that electronic submission of manuscripts has become routine, the requirements of many journals have become even more detailed. In the case of a research paper, for example, there may be editorial requirements about the number of manuscript copies needed; the paper quality and typeface; the margins and spacing; the contents of the title page; the methods used for documentation; the form in which figures, tables, and other visual materials are to be prepared; and so on. Prospective authors who fail to follow the submissions guidelines of the journal for which they are writing virtually guarantee that their manuscript will receive little serious attention.

For academic assignments, too, a neatly prepared manuscript suggests that you are serious about your work and have devoted care and thought to all stages of its preparation. Check if your instructor has his or her own format specifications. Otherwise, the guidelines below will help you produce a professional-looking manuscript.

Paper, Margins, and Spacing

Use 8½-by-11-inch white paper and a standard typeface (not script type or all capitals). A font size of 11 or 12 points is typical. Print on only one side of the paper. Leave margins of 1 to 1½ inches on all sides of the page, and do not justify the right margin. Indent each new paragraph one-half inch or five spaces.

Biologists submitting manuscripts for publication are usually asked by journal editors to double-space the entire manuscript, including the Abstract and the Literature Cited section. This format is preferable, too, for academic assignments. Another option for academic assignments is to use single-spacing within each entry in the Literature Cited section, with double-spacing between entries; the Abstract may also be single-spaced. Check with your instructor to see if he or she has spacing preferences.

Journals vary with respect to whether indentation is used in the Literature Cited section. In academic papers, a common practice is to use a hanging indent, placing the first line of each entry flush with the left margin, and indenting successive lines of the entry one-half inch (or three to five spaces).

Title Page

An acceptable format for a student paper calls for the following information on the title page: title of the paper, your name, course information, and the date. Center the title and place it about one-third down the page; center and double-space all successive lines under the title. It is not necessary to underline the title, to put it in quotation marks, or to follow it with a period. Capitalize the first letter of all important words on the title page—that is, words other than articles (*a, an, the*), coordinating conjunctions (*and, but, or, for, nor, so, yet*), and prepositions (*in, on, to,* and the like).

Pagination

Number pages consecutively beginning with the title page (which does not actually carry a number, but is still counted). Use only arabic numerals (1, 2, 3, . . .), and put the number in the upper right-hand corner of each page. Writers submitting a research paper for publication often begin each major section of the paper (Abstract, Introduction, Materials and Methods, and so on) on a separate page. This is not necessary for a student paper, although you may find that separating the sections makes a long paper easier to follow.

Headings

Headings are useful organizational tools. They can designate major portions of a work (for example, the Introduction or the Discussion of a research paper), and they can be used in a review paper to divide a long or complex text into smaller, more coherent, and more manageable sections. The following guidelines will help you use headings correctly.

1. Use headings and subheadings sparingly. Student papers generally need no more than one or two levels of headings. Overuse of headings results in a fragmented text, with many small sections. As a result, readers may have difficulty following your organizational logic and seeing how the various sections fit together.

2. Make headings informative and concise. Headings function like signposts, helping to orient readers. They should not be cumbersome and tedious to read.

3. Use the same format for headings on the same level of organization. For example, you may decide to center and capitalize all first-order headings and to underline second-order headings at the left-hand margin:

```
                    RESULTS

Outcrossing rate

As shown in Figure 1, the proportion of spores

that formed appressoria at the end of germ

tubes . . .
```

Whatever you decide, however, be consistent or else you will confuse your readers.

4. Use parallel grammatical form for headings on the same level of organization.

For example, you may wish to use all nouns, or all verb forms ending in *-ing*. Whatever grammatical form you choose, be consistent. For example, suppose that your first two headings in the Results section of a lab report were *Mode of infection* and *Temporal variation of infection*. Subsequent headings should be parallel: not *Measuring spatial variations of infections,* but *Spatial variations of infections, Consequences of infection,* and so on.

Figures and Tables

Biological journals usually require authors to put each table or figure on a separate page. The title of a table is typed on the same page as the table; however, figure legends are typed together on a separate page. Tables, figures, and figure legends are then placed at the end of the Literature

Cited section in that order. For electronic submission of figures, journals have additional technical specifications.

You may wish to follow the above procedure for academic assignments, unless your instructor prefers that pages containing tables and figures be incorporated with pages containing text. If so, put each table or figure after the page on which it is first mentioned and number all pages. In this case, each figure should be accompanied by its legend, which can be typed on the same page.

Still another way to deal with figures and tables is to insert them directly into your text pages, again putting each figure or table as close as possible to the point at which it is first mentioned. This will make your paper resemble a published work. Most good word processors can import graphics into a text file, allowing you to position a figure or table on a particular page, with writing either above or below it.

Assembling the Manuscript

Never submit a loose pile of pages: use a staple or a paper clip to fasten everything together. A special folder is not necessary (although it's a nice touch). In case your instructor misplaces the paper, keep a copy of the original on your computer or make a photocopy of the original. You should also make a backup copy of your files on a CD-ROM or some other external medium.

WRITING AN ACKNOWLEDGMENTS SECTION

In published biological papers, a short Acknowledgments section usually comes between the Discussion and the Literature Cited sections. Here you express thanks to people who assisted you with the research itself or with the preparation of the paper. Scientific ethics dictate that you first obtain the permission of all those you intend to acknowledge, preferably by showing them a copy of that section of the paper. This is because it is conceivable (though probably unlikely) that some people who offered you help may not wish to be associated in print with the paper or may object to the wording you used to refer to them. The Acknowledgments is also the place to list the sources of any financial support you received.

You may wish to include an Acknowledgments section at the end of some course assignments, such as an honors thesis or senior independent research project. Whether in published or unpublished works, the Acknowledgments should be brief and concise. Write in full sentences, not fragments, and avoid informality and slang. Keep your tone consistent with that of the rest of the paper. Refer to people by their full names (or first initials and last name, depending on journal style), and omit titles.

Poor	I was supported while doing this research by summer money from Astor College. Thank you so much, Professor Huerta, for spending hours helping me do the literature review for this paper. And thanks to Max for showing me how to analyze the data. Gus Arnold also gave me suggestions on how to revise this paper. Finally, thanks to Mallory and Giles for their words of encouragement and for all that great pizza!
Improved	This research was supported by a summer research stipend from Astor College. I thank Deborah Huerta for help with the literature review, Max Rider for assistance with data analysis, and Gus Arnold for useful comments on an earlier draft of this paper.

In the revised version, the writer has replaced fragments with full sentences and now uses first and last names without titles. She has given more specific information about her funding and has deleted unnecessary wording. The pizza might have been delicious, but it doesn't need to be mentioned in the Acknowledgments.

See also the sample papers in Chapters 4 and 5 for examples of Acknowledgments.

PROOFREADING

Typographical errors, misspelled words, missing commas or periods, irregular spacing, and other minor errors distract the reader and undermine your authority as a writer. The aim of proofreading is to eliminate such mistakes from the final draft of the manuscript.

It is not easy to proofread well. By the time you have worked through several drafts of the paper you are no longer as observant as you might be. You will tend to read what you *think* is there, not what is actually on the page. Therefore you must force yourself to look at every word, space, line, punctuation mark, and number, including all material in figures and tables. Proofread the whole manuscript more than once over a period of several hours or days; you may miss some errors the first time but notice them during the second or third proofreading, when you have achieved more distance from the material. Ask a friend to look over the paper as well.

Finally, learn to use word processing features that speed up revising and proofreading (see pp. 171–172).

CHAPTER 9

Using Writing to Prepare for Examinations

It is widely recognized that writing facilitates learning. Conversely, if you do not fully understand something, you will probably have difficulty writing about it. Often the process of finding the right words can help correct flaws in your reasoning and clarify your thinking. Even the *mechanical* act of writing, whether by hand or on the computer, gets you more involved with the material. Writing *as* you read or think reinforces your memory of the facts and can reveal connections between related ideas.

GETTING THE MOST OUT OF TEXTBOOKS

Following are several ways to make reading your course texts an *active,* not a *passive,* experience, and thus better prepare you for labs and exams.

■ Use highlighting sparingly.

You may be used to emphasizing important parts of a text by underlining them or using a highlighter pen. This is a useful reading strategy but only if used thoughtfully. Avoid highlighting line for line *as* you are reading; it doesn't allow you to make meaningful distinctions about the relative importance of successive sentences. It is better to *first* read a passage all the way through; then go over it again and highlight *selective* points. The reading will take longer, but you will learn more this way.

■ Annotate your textbooks.

Adding your own annotations—handwritten questions, notes, and other comments in the margins—is another way to read interactively.

Make difficult passages comprehensible by summing them up in your own words. Insert questions next to material you don't understand. If certain points relate to lectures or lab exercises, make notes to that effect so that you can integrate various components of the course. Used selectively with highlighting, annotation can individualize your texts and make information easier to access and retain.

■ Read chapter summaries and respond to review questions.

These study aids reflect the author's expert insights about the most important material. If your text does not have a chapter summary, write your own. If a summary is provided, try composing your own version anyway. Write out the answers to end-of-chapter questions, even ones you think you understand; do not simply answer them in your head.

TAKING GOOD LECTURE NOTES

Over the years, you have probably developed your own strategies for note taking. By now you know that good class notes can be invaluable for self-teaching. The following suggestions will help you use writing more effectively to study biology and many other subjects.

■ Be selective.

During lectures, do not feel you have to record every word. If you write continuously, you will not be giving full attention to what the speaker is saying. Your notes may be voluminous, but they may not make sense because you won't be thinking about what you are writing. Taking notes nonstop also inhibits your ability to ask questions about what you do not understand and to contribute to class discussions.

Good lecture notes are a selective record. They reflect the writer's attempts to summarize and synthesize the most important points, to reproduce the logic and continuity of the speaker's discussion, and to make links to material from previous class meetings. Many of your notes may be abbreviated versions of the lecturer's own sentences, but others should be in your own words, particularly when you need to bridge the gap between one topic and another.

■ Revise your notes as soon as possible.

You probably use various shortcuts to take notes quickly, such as writing in sentence fragments and abbreviating words. Your writing may make sense at the time but be hard to decipher later. For this reason, review each day's notes while the material is still fresh in your mind. Write out any shortened sentences or words that may confuse you later. Clarify difficult concepts by adding comments in the margins. Check that diagrams are accurately labeled and integrated with the rest of your notes. Number related items and use underlining, asterisks, or other labels to call attention to the most important ideas. Reworking your notes after each lecture may seem time-consuming, but it will reward you later when you are studying for examinations. Moreover, it reinforces your understanding of the material.

■ Keep up with assigned readings.

Instructors write their lectures assuming students have acquired certain background knowledge by reading the textbook or other course materials. If you come to class with that knowledge, you are more likely to understand the lecture and to take coherent notes. Moreover, you will be less likely to waste time and energy writing down information covered adequately elsewhere.

PREPARING FOR LABORATORY EXAMS

Introductory and broad survey courses in biology often have laboratory exams, or practicals, as well as lecture examinations. Preparing for a lab exam may be difficult if you have not taken good notes during the laboratory periods. Much of the advice for lecture notes also applies here. Remember the benefit of advance preparation: read both your text and your lab manual and understand how each lab's exercise fits into the framework of the course as a whole. Record observations and questions as you work; do not trust your memory.

When preparing drawings, use your text or other references to help you identify structures, but draw what you *see;* do not simply copy a stylized textbook illustration. Suppose, for instance, you wish to draw a cross-section of a *Smilax* root from a microscope slide (see Fig. 9.1 and the sample student drawings based on it in Fig. 9.2). You need not be an artist to render an accurate, detailed, and comprehensible illustration. Use pencil for your initial sketch and allow enough room to include sufficient detail. Label your diagram carefully and always print legibly; be sure to identify details using the correct terminology. Check that all lines pointing to particular structures are drawn unambiguously. Provide a legend so that you

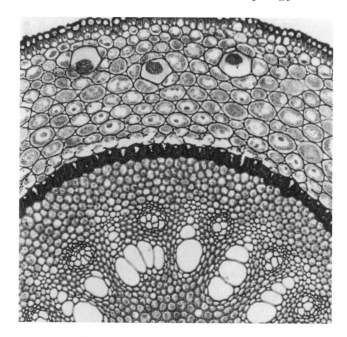

FIGURE 9.1 Photomicrograph of *Smilax* root in cross-section (100 X)

POOR

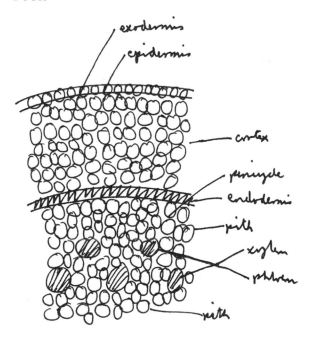

FIGURE 9.2 *Poor* and *improved* sample student drawings of *Smilax* root based on Figure 9.1 (cont'd. on p. 210)

IMPROVED

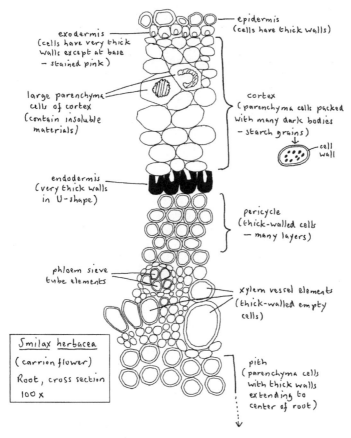

epidermis
(cells have thick walls)

exodermis
(cells have very thick
walls except at base
— stained pink)

large parenchyma
cells of cortex
(contain insoluble
materials)

cortex
(parenchyma cells packed
with many dark bodies
— starch grains)

cell
wall

endodermis
(very thick walls
in U-shape)

pericycle
(thick-walled cells
— many layers)

phloem sieve
tube elements

xylem vessel elements
(thick-walled empty
cells)

Smilax herbacea
(carrion flower)

Root, cross section

100 x

pith
(parenchyma cells
with thick walls
extending to
center of root)

FIGURE 9.2 (continued)

know exactly what is being represented; note the magnification for microscope drawings.

Soon after each lab period, go through your notebook, expanding it with additional comments and redrawing illustrations that might confuse you later. Finally, if possible, take full advantage of review labs. See Chapter 2 for additional tips on recording information in the laboratory and the field.

STUDYING FOR SHORT-ANSWER QUESTIONS

Most biology exams include a variety of short-answer questions. For example, there may be multiple-choice, matching, or true/false sections, as well as questions calling for answers of a few words to a few sentences. Do not simply reread your notes and textbook to study for such tests. Following are ways to use writing productively to make studying a more active experience.

■ Create your own expanded "textbook."

Make one coherent study guide by integrating lecture notes, course readings, and material from laboratory or field sessions. After all, your instructor sees all these components as related to one another; you should, too.

■ Compile a list of key terms.

Look for words appearing in italics or boldface in your text, and check the end of each chapter for a glossary or list of important words.

Also, consult your notes for terms that your instructor has emphasized or used repeatedly. Write out the definition of each term in your own words; then check your book or notes for accuracy. Do this repeatedly until you have mastered the important vocabulary.

■ Devise your *own* questions.

Then devise "perfect" answers. Index cards are useful here: you can put the question on one side and the answer on the other. Don't simply flip the cards; write out your response to each question again on a separate sheet of paper and check it against your prepared answer. Repeat the process until you feel confident with the material. This activity is also an excellent strategy for studying for essay exams (see below).

■ Do not become lost in details: focus on central issues.

One of the most difficult tasks for beginning students is deciding at what level to aim their mastery of a subject. In practical terms, this translates into worrying about how picky an instructor is going to be on tests. This problem has no simple solution. However, even when you are aiming for quick recall of a large body of information, focus first on the main points. Biology is not a collection of isolated facts: much more important are the underlying *patterns* and unifying *concepts*. Try grouping related items: three factors that influence a particular process; major similarities between two groups of organisms; six characteristics of a particular family of animals; and so on. Also consider the interrelationships that may exist among important ideas. Some people find it helpful to diagram these connections, starting with a single main idea or topic and surrounding it with related concepts joined by connecting lines. You will remember the particulars if you first understand the generalities.

■ Find a study partner, or create a study group.

Research has shown that many people profit from collaborative learning. Consider working with one or more compatible, motivated students on a regular basis to revise and review lecture notes and prepare for exams.

ANSWERING ESSAY QUESTIONS

Successful essays on examinations are usually the product of much advance preparation. If your studying has been thorough and well focused, you will have tried to make connections between topics, identify central issues, and see the course material in new ways. All of these skills are usually tested by essay questions.

The advice on preparing for short-answer questions applies as well to essay exams. However, essay questions also pose other challenges. Awkward, wordy, or ambiguous prose will suggest that your thinking is also sloppy and unrefined and that you do not fully understand the material. Clarity and coherence are as important in essay exams as in other forms of biological writing, yet time constraints usually do not allow much opportunity for rewriting. For this reason, good essays are the product of well-focused studying followed by careful use of time during the examination.

The following strategies may help you.

■ Read through the entire test before you start.

Check whether you must answer all questions or if you have some choice. Do the *easiest* sections first. Budget your time carefully; give yourself a time limit for each question. Leave a few minutes at the end to review your answers and to expand them if necessary. Check again that you have answered or attempted all required questions.

■ Address the question *asked*.

This may sound obvious; however, many students do not read the question carefully. Some essays call for a straightforward summary of information, whereas others ask you to think more creatively. Make sure you understand this difference. Focus on key words in the question: *define, list, compare, contrast, evaluate, analyze, describe, discuss*. Look for possible choice *within* questions. For example, you might be asked to define four terms out of five or to answer one part of a question from a choice of three parts.

■ Stick to the point.

You will not get a higher grade for adding irrelevant information. Such material obscures the strong parts of your answer and suggests that you don't fully understand the subject. It also wastes valuable time.

■ Observe length restrictions on answers.

Guidelines for the length of answers (one concise paragraph, two pages of the examination booklet, and so on) convey your instructor's advice about the depth of detail needed and about the amount of time you should reasonably spend on the question. Point values for specific questions also guide you in this respect. If a question is worth 10 points out of 100, avoid spending more than 10% of your time on it—in other words, no more than 5 minutes during a 50-minute exam.

■ Plan before you start to write.

Organize your thoughts by jotting down a brief outline or list of important points. Crucial facts are easy to forget when you are working under pressure. Taking a minute or two to arrange your thoughts will save time later and help you write a more complete answer.

■ Develop a clear thesis.

Essays requiring you to integrate, analyze, or evaluate material can easily lose focus. Do not ramble; organize your answer around a clearly stated central idea. State this in a topic sentence (see p. 176); then build the rest of the essay around your main point. Sometimes the wording of the question gives you raw material for a thesis statement; however, don't waste time unnecessarily repeating the entire question.

■ Support general statements with specific evidence or examples.

Many essays are inadequate because the writer conveys only a superficial knowledge of the material. Thorough understanding of a subject involves the ability to explain how broad ideas rest on a foundation of supporting details. Here is your chance to display the depth of your preparation.

■ Include illustrations, if appropriate.

Sometimes examination questions call for specific drawings or graphs. On other occasions, you may choose to add these features to supplement your discussion. Be sure that all your illustrations are accurately labeled and large enough to convey information clearly; otherwise, they may actually detract from your answer. Do not simply tack on drawings next to your text; instead, refer to them specifically as you write.

■ Sample answer to an essay question

Following is an example of an answer to a short essay question on an invertebrate zoology exam. The question reads as follows: Briefly describe the process by which a blood clot forms, and explain how the saliva of the medicinal leech (*Hirudo medicinalis*) specifically interferes with this process.

POOR   *The process by which the saliva of the medicinal leech interferes with blood clotting is that in leech saliva there is an anesthetic as well as a blood clotting inhibitor. The saliva of a medicinal leech contains hirudin, which will prevent blood clots. Another species of leech contains hementin, which can destroy clots that are already formed. Leeches are now used in medicine to help keep blood flowing after certain kinds of surgery. The medicinal leech produces a substance that interferes with the production of the proteins needed in blood clotting.*

This answer has several problems. First, the student wastes words and valuable time by repeating the question, or part of it. Second, she does not completely address the question, since she fails to describe the process by which a blood clot forms. Third, she wastes time on irrelevant information (medicinal leech saliva contains an anesthetic; leeches are used after certain kinds of surgery; a different species of leech contains hementin). Finally, she does not fully answer the last part of the question: how, specifically, does the medicinal leech interfere with clotting? Her answer suggests that she does not understand the material and is simply writing down whatever stray facts she can recall.

Now consider another possible answer to the same exam question:

IMPROVED   *When a blood vessel is injured, clotting factors are released from platelets and damaged cells. These factors help to promote the conversion of prothrombin (a protein in plasma) to thrombin. Thrombin is an enzyme needed for the conversion of fibrinogen, another plasma protein, into fibrin, still another protein. The threadlike strands of fibrin trap blood cells, producing a clot. The saliva of the medicinal leech contains an anticoagulant, hirudin, that blocks the conversion of prothrombin to thrombin, thus interfering with clotting.*

This student has grasped the intent and scope of the question. She understands the material in some detail and has conveyed this knowledge clearly and coherently. Her answer addresses both parts of the question and includes no irrelevant information.

CHAPTER **10**

Other Forms of Biological Writing

ORAL PRESENTATIONS

Many biologists participate in professional conferences or symposia that offer a forum for exchanging ideas, conveying information about new research methods, and reporting research in progress. Conference proceedings are often available in printed form; however, oral exchange of information is still an important means of communication.

As a biology student, you may be required to prepare an oral presentation of your own—perhaps a summary of a published paper, a review of the literature on a particular topic, or even a report of your own research for a seminar or a professional conference. Although the final product will not be presented in written form, you will still need to use writing to organize your talk. Actually, there are many similarities between writing a paper and planning an oral presentation. Both activities require you to understand your audience and your purpose and to convey information clearly, accurately, and logically. Both also force you to examine your own understanding of the material and to use writing as a means of clarifying your thoughts.

Following are some practical suggestions for giving successful oral presentations.

■ Consider your audience.

Tailor your presentation to your *listeners.* Consider their experiences, interests, and background knowledge in the context of your own rhetorical aims. For example, you would pitch your talk differently to a

215

nonscientific audience than to a scientific one. For a lay audience, you might choose a different title, use fewer technical terms, cover the material in less depth, omit certain aspects of your topic, even take a completely different slant. Also, of course, audiences have different expectations depending on the purpose of your talk in the first place. A course instructor will expect you to demonstrate sound knowledge of your subject and the ability to handle questions from the rest of the class. Conference participants will be interested in the quality of your research, as well as its implications for their own work.

■ Use an appropriate method of organization.

How you organize the presentation will depend, of course, on the subject, your audience, and your objectives. Nevertheless, planning will be easier if you think of the talk as having a distinct beginning, middle, and end. A logical way to start is with general background material; then narrow down to the specific focus of your talk. Be sure to keep introductory material brief, or else you'll have to rush through the most important points. Save time at the end to summarize, offer conclusions, discuss broader aspects of the topic, and answer questions.

If you are reporting original research, use the format of a research paper (see Chapter 4) to guide you. Begin by briefly summarizing the general topic or scientific issue, placing your own study in context. Then get specific: Why did you do this study? What did you expect to find? Briefly explain your materials and methods. Next, in the middle (and longest) section of your presentation, summarize the results. Here, focus on your major findings, adding just enough supporting information (details, statistics, examples) to develop your points. Conclude by looking again at your objectives. Do the data support your original hypothesis? What connections can you make to the findings of other researchers? What questions remain? It may also be appropriate to acknowledge specific people who have helped you with the research, as well as any sources of funding. Some speakers do so at the beginning of their talk, others at the end.

■ Write out the entire talk beforehand.

Even if you are an accomplished speaker, putting everything down in writing will make your presentation more organized and coherent and will lessen the chance that you'll forget an important point. Plotting out the talk sentence by sentence also allows you to plan your words carefully and search for the most effective ways of explaining difficult ideas. After you rough out the first draft, revise the manuscript carefully, looking for sections that are poorly worded or likely to confuse the audience or that jump abruptly from one topic to another.

■ Never *read* a prepared talk word for word.

Doing so will suggest (perhaps correctly) that you are not comfortable enough with the material to abandon your notes. It will also distance you from your listeners, who will rapidly lose interest in your formal recitation. Instead, use the written version of the talk to make a brief list of key points or concepts; these can be put on a single sheet of paper or on file cards arranged sequentially. As you speak, use the key points to jog your memory and keep you on track. Once you become thoroughly familiar with the talk in its written form, you will probably remember your most effective sentences and phrases word for word and will be able to say these naturally as if they just occurred to you. An effective talk—one that really engages the audience—strikes the proper balance between carefully structured wording, worked out in advance, and a spontaneous, informal delivery.

■ Observe your time limit.

Presentations at professional conferences are subject to strict time constraints: if you run out of time, you may have to stop abruptly even if you're only midway through your talk. Even student presentations must take into account the instructor's plans, the needs of other speakers, and the duration of the class period. Beginning speakers often devote too much time to introductory material or their first few points and then run out of time at the end. Time your presentation carefully, either by running through it mentally or (preferably) rehearsing it out loud. Decide what you should be discussing halfway through the allotted time period, and note this point both on the final draft of the talk and on your speaking notes. Running over time is not only a sign of poor preparation but also a discourtesy to others. The same applies to starting late.

■ Do not speak too rapidly.

If you are nervous you may speak more quickly than usual, especially as you approach the end of the talk. Remember that listeners need time to digest everything they hear. If you confuse them at any point, they may stay confused for the rest of the talk; unlike readers of written text, they cannot go back and review difficult sentences. It helps to pause briefly after important points and to repeat difficult material in slightly different wording. If your format permits, you can also invite questions at potentially confusing places in the presentation.

■ Do not swamp listeners with details; avoid jargon.

Develop only a few main points. Even highly attentive listeners can take in only so much information at one time; they'll lose track of your argument if you bombard them with too many details. Use clear, straightfoward language and avoid jargon. (For more on writing clear, accurate sentences, see Chapter 7, pp. 178–187.) Explain any terms likely to be unfamiliar to listeners, but keep the number of such terms to a minimum.

■ **Establish eye contact with the audience.**

Doing so will make you more relaxed and your audience more receptive. Addressing your listeners directly also encourages you to speak up, not mumble as you gaze at the floor. Your voice needs to be loud enough to capture everyone's attention, even those in the back row.

■ **Use visual aids.**

PowerPoint

PowerPoint (Microsoft Corp.) is a versatile software package that allows you to mix text with graphs, tables, photographs, or videos, including animation. With PowerPoint you can prepare separate slides, shown via a computer projection system, that lead your audience through your presentation from start to finish. The basics of PowerPoint are easy to master; however, don't let the technology overshadow the content of your talk. Even if you have mastered all the bells and whistles, if you have little to say, your presentation will still fall flat. In fact, overuse of every feature and gimmick may distract, even irritate, your listeners. Following are some guidelines advocated by many PowerPoint users.

Each slide should focus on a single topic. Don't use more text than your audience can take in at a glance—a general rule of thumb is to use a maximum of 5–6 lines per slide and no more than 5–6 words per line. Try not to break words at the end of a line. Align type on the left, but not on the right, and use sans serif fonts, such as Arial or Verdana, for more readable text and graphics. Some speakers prefer serif fonts for titles. Avoid using all capitals, even in titles, since such text is difficult to read. Proofread all text meticulously for misspelled words and typos, which will be embarrassing later when they are projected on a big screen.

Remember that some people will be looking at your slides from the back of the room, perhaps a large room, so your type size should take this into account. Use type of at least 36 points for titles and at least 24 points for the text. If necessary, you can add different colors, as well as italic or bold, for emphasis, but use these devices sparingly and consistently.

Bullets can be effective in setting off related points in a list. For consistency, use parallel grammatical construction for all the bullets on a slide. For example, if three of your bulleted points are sentence fragments beginning with -*ing* forms of verbs, then your fourth point shouldn't be a full sentence beginning with a noun. Punctuation is not necessary at the end of a bulleted point.

Bear in mind the importance of contrast in creating readable slides. For instance, you might use light letters on a dark blue background. For an integrated effect, keep the same background color from slide to slide, and don't overuse color (or any other attention-getting device).

Do not simply read your text slides back to the audience word for word or you will lose your credibility — and your listeners. The audience still expects you to give a coherent talk in your own words. Your slides should highlight your main points, not substitute for your presence.

Slides depicting numerical data should be simple enough for the audience to grasp quickly. Do not pack them with more information than people can absorb in a minute or two. Axes of graphs should be clearly labeled and, as with text, all words and numbers should be large enough to be seen from the back of the room. Remember that slides showing data should stress concepts, not particulars. Use them carefully and sparingly to make your points.

For original research, you may wish to incorporate digital photographs illustrating the field site, organisms studied, experimental apparatus, etc. These pictures must be sharply focused and well composed or else they are not worth using. Do not make the common mistake, seen even at professional meetings, of showing blurry, cluttered, or otherwise unsuitable photographs accompanied by weak apologies ("This picture isn't that clear, but . . ."). If you are not a competent photographer, ask someone for help, or think of an alternative method to convey the same information. Clip art can be useful in PowerPoint presentations, but only if used sparingly or else it will distract from your content.

You may wish to bring along supplementary handouts—additional references perhaps, a more detailed account of your methods, or reprints of an earlier study. These can be made available later to anyone who is interested.

SAMPLE POWERPOINT PRESENTATION

Following are slides for a PowerPoint presentation of the student research project in Chapter 4 (pp. 102–113). Notice how the speaker has focused on the most important aspects of the study, using slides to highlight these points rather than simply present a condensed version of the paper.

Other visual aids

Many speakers still rely on 35mm transparencies to illustrate their talks. For a conventional slide presentation, most of the same guidelines for PowerPoint still apply. Transparencies should be sharp, uncluttered, and well-exposed; otherwise, don't use them at all. Decide beforehand exactly what you will say about each slide, and omit any slides that serve no clear

Germination of *Arabidopsis thaliana* in Response to Gibberellins in Light and Darkness

Kristin A. VanderPloeg
Dept. of Biology
Colgate University

Slide 1

Gibberellins and Seed Germination

- Break seed dormancy
- Stimulate germination
- Promote embryo development

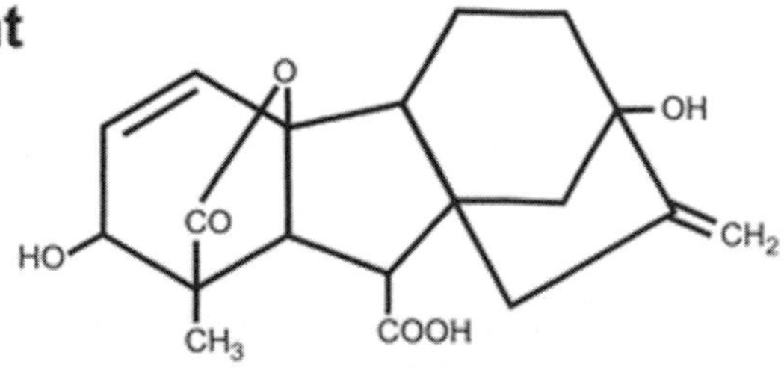

Gibberellic acid (GA_3)

Slide 2

Study species

- *gal-3* mutant of *Arabidopsis thaliana*
- Germination depends on exogenous GA's

Questions

- How does GA_3 concentration affect germination?
- Are effects of GA_3 different in light *vs.* darkness?

Slide 3

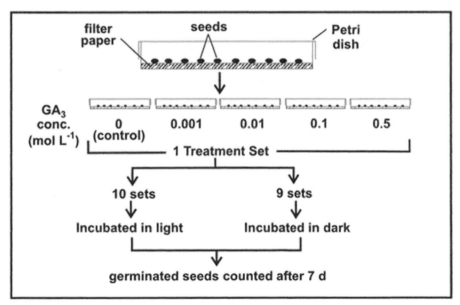

Slide 4

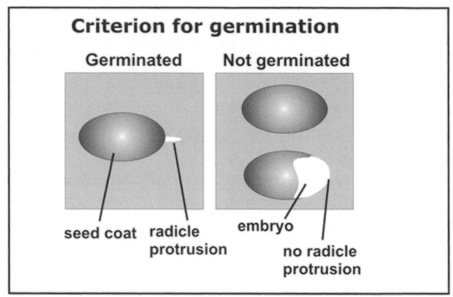

Slide 5

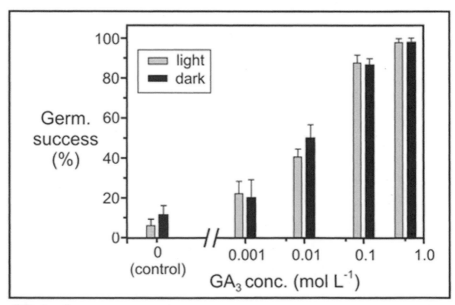

Slide 6

Two-way ANOVA: GA_3 conc. and light/dark conditions

Source of variation	F
GA_3 conc.	105.96***
Light/Dark	0.41^{ns}
GA_3 conc. x Light/Dark	0.48^{ns}

*** = $P < 0.001$; *ns* = nonsignificant ($P \geq 0.05$)

Slide 7

Conclusions

- Germination success increased with increasing GA_3 concentration in both light and darkness

- At all GA_3 concentrations, germination success did not differ in light *vs.* darkness

Slide 8

purpose. Write down in your lecture notes when each slide should be shown. Budget your time carefully to take into account how much time you will devote to each slide. When possible, try to show slides in one or more groups; otherwise, you will distract the audience by repeatedly turning the lights on and off. Set up the projector and screen in advance, locate a pointer, and check that all slides are loaded correctly and in the proper order.

Overhead transparencies are also effective lecture illustrations. You can use blank transparency sheets to jot down terms or make quick drawings as you talk. Even better, you can prepare transparencies of bulleted lists, graphs, tables, etc., in advance. Again, keep these simple, concise, and easy to read. An advantage of overhead projection is that it allows the speaker to leave the room lights on and thus to interact more directly with the audience.

Finally, for classroom presentations, the blackboard is a traditional and still invaluable tool. Use it as you talk to write down unfamiliar terms or important statistics or to make very simple drawings or graphs. Your writing should be large enough to be seen clearly from the back of the room. Usually printing is easier to read than script. Graphs should be clearly labeled and easy to grasp quickly. Do not clutter the board with disorganized scrawls, and do not inadvertently erase information you must refer to frequently.

Material that is relatively time-consuming to draw should be put on the blackboard *before* your talk; otherwise you'll waste time writing while listeners wait impatiently or struggle to copy what you have written. Consider, instead, putting such material in a handout that the audience can look at as you talk. Handouts are also useful for listing key terms and definitions, important points to be covered, or useful references. However, don't simply read your handouts aloud to your audience, and don't make them too detailed; otherwise, people will spend their time reading instead of listening to you.

■ Rehearse your presentation in advance.

If possible, find several caring people who will listen carefully to your talk and give constructive criticism. Ideally, these should be people whose background and interests are similar to those of your "real" audience. Make this practice session a full rehearsal, complete with slides, overhead transparencies, etc. If possible, rehearse in the room in which you'll actually be speaking or in one of similar size and layout. Have at least one person sit at the very back of the room. Are you talking loudly enough? Are you speaking clearly, neither too fast nor too slowly, and with enough energy so as not to be tedious to your listeners? Are all slides or transparencies readable? Also ask one of your listeners to time your talk from start to finish, allowing a few minutes for questions and answers at the end. Rehearsing in front of a live audience will not only allow you to fine-tune your whole presentation, but it may also help you feel more confident later.

■ Be prepared for questions.

You cannot predict everything you will be asked, but you probably can anticipate some of the questions. Write them out beforehand and prepare brief, concise answers. It often helps to repeat or rephrase each question before you answer it, since some people may not speak loudly or clearly enough for others to hear. Doing so will also give you a little extra time to compose your response.

If you are asked a question for which you are unprepared, do not try to bluff your way through a reply. It is far better to say that you don't know the answer. If you have given a thoughtful, well-organized talk, listeners will already be convinced that you know your subject. They will not expect you to know everything.

■ Check out the facilities and equipment in advance.

Regardless of what kinds of visual aids you plan to use, be sure to double-check the facilities, especially if you will be speaking in an unfamiliar setting. Many conference rooms have blackboards and overhead projectors, for example, but some do not. Pointers and microphones may or may not be part of the standard equipment, and important light switches may not be located in obvious places. If you plan to use Power-Point for a conference presentation, you may need to follow specific guidelines about submitting your material. Even for a talk at your own institution, it is vital to check out the computer projection system you will be using. As additional insurance, prepare overhead transparencies as a backup to use if you run into technical problems.

In summary, it is wise to arrive early and check the setup well in advance. Doing so will help you feel more comfortable and help you focus on the most important part of your talk: what you actually want to say.

■ Checklist for oral presentations

- Are you sure of your own objectives as speaker?
- Do you understand the expectations and background of your audience?
- Have you prepared a written version of your talk? Do you have a clear organizational plan, with a beginning, a middle, and an end?
- Have you emphasized main points and trends? Do you save time for a summary and conclusions?
- Do you have brief lecture notes to consult if needed?

- Have you prepared high-quality visual aids (e.g., PowerPoint, 35-mm slides, overhead transparencies) to illustrate and emphasize your points?

- Have you checked out the room and equipment in advance?

- Do you have backup materials in the event of technical problems?

- Are you prepared to speak spontaneously, without reading your notes (or your overheads or slides)?

- Have you rehearsed your talk before a live audience?

- Is the length of your presentation within the time limits?

- Are you prepared for questions?

POSTER PRESENTATIONS

A poster presentation conveys an author's original, unpublished findings *visually* through a selective assemblage of illustrations (graphs, tables, drawings, photographs, etc.) that are carefully integrated with a small amount of text. Thus, an entire study can be summarized visually. Posters have become popular forms of communication at scientific conferences, and considerable space may be set aside for them. Usually, there are designated time periods for the authors to be present to answer questions. Some people prefer giving posters to delivering talks because they feel more comfortable with the format; in other cases, a particular study may lend itself more readily to a poster presentation.

From the audience's perspective, browsing through a poster session at a busy conference offers a pleasant interlude during a tightly scheduled sequence of talks. In addition, because oral presentations are generally brief, with an even shorter question-and-answer period, poster sessions allow for more extended, informal discussions between a researcher and those interested in his or her work. Finally, poster sessions offer a practical solution to the problem of fitting in a large number of presentations during a limited time period.

If you are preparing a poster presentation of your own study, the following advice may help you get started.

■ Follow guidelines closely.

This advice applies whether your poster is for a course assignment, a departmental poster session at your college, or a professional meeting. One conventional way to produce a poster is to mount sections of text along with graphics onto a large piece of mat board or other backing material. Alternatively, you may have access to equipment that enables you to print out the entire poster on a single large sheet of paper that can then be mounted onto a firm surface. Whatever construction method you use, be sure you understand all details about poster size and contents, the nature of the display surface (easels? bulletin boards?), and the availability of thumbtacks, tape, or other materials. Requirements and facilities for poster presentations can vary widely.

Remember, too, that poster presentations for a conference, like oral presentations, must be formally accepted by the organizing or editorial committee of the conference. Prospective presenters are usually asked to submit a title and an abstract. It makes little sense to spend large amounts of time preparing a poster until you know it has been accepted.

■ Remember your audience.

Like oral presentations, a poster presentation should be geared to its intended audience. If your viewers will be specialists within your narrow field, then assume they have some background knowledge and will understand most technical terms as well as the basic scientific issues. They will probably be interested in how your work relates to their own. If your poster is for a course-based or departmental presentation, then most viewers will be your classmates, many of whom will have backgrounds similar to your own. If your work is intended for a heterogeneous or nonscientific audience, then you may wish to reassess your diction and general approach, perhaps taking a slightly different slant.

■ Use a simple and logical organization.

Scientific posters typically follow the basic plan of a research paper. Start with a concise, specific, and informative title. Next, introduce the scientific problem and your specific research goals, question, or hypothesis. Describe your materials and methods; then present your data using figures and/or tables along with concise text. Add a brief discussion or conclusions section, interpreting results in the context of your objectives or hypothesis and with reference to other studies. Alternatively, you may wish to combine results and discussion to save words and space. End with a brief Cited References section and with Acknowledgments if needed.

Some authors of posters also include an Abstract, placed before the introduction. However, many people omit this section to avoid needless repetition and allow more space for other material. Moreover, if your Abstract has already been included in the conference bulletin, then viewers of your poster do not need to see it again. Check conference guidelines or ask your instructor about his or her preferences.

■ Choose an effective layout.

As a mainly visual document, a poster should be engaging, organized, and uncluttered. There is no one right way to design a poster, but some ways are more effective than others. If possible, study examples of other people's posters before you make your own. Decide for yourself what works well and what doesn't. Take notes on successful designs and formats, and adapt the best for your own use.

Remember, too, the way most people read. Your layout should invite readers to start at one place and proceed logically according to a particular sequence. The title of your poster, which should run along the very top, is in a position to attract (or deter) readers, so make it both informative and engaging. A common design for horizontally arranged posters is to use several columns, with the introduction at the top left and the last section (literature cited or perhaps acknowledgments) at the lower right. Each column then reads from top to bottom, and the entire poster from left to right.

Some people insert arrows to help lead the eye from one section of the poster to the next. Bullets or numbers are also useful to set out lists of points.

Finally, remember the visual function of blank space to draw attention to separate portions of the poster and also to rest the eye.

■ Keep text to a minimum. Eliminate all extraneous material.

Probably the most common mistake made by authors of posters is to include too much text. Viewers should be able to grasp the main thrust of your work in just a few minutes. However, you'll need to attract and hold their attention in the first place. Most people don't have the patience, time, or interest to stand around reading posters that resemble entire research papers simply pasted onto backing material. Such posters usually receive few visitors.

Resist the temptation to include every point, every sentence in your original paper, no matter how difficult it was to compose. Keep background material brief and get right to the point. Prefer active over passive voice. Pare down your methods and materials to the bare essentials — many authors use flowcharts, drawings, photographs, etc., to simplify this section. Present only your *most important* results; save the rest for informal discussion during the poster session. Consider using lists of key points (these can even be sentence fragments) to replace paragraphs of narrative.

Similarly, think carefully about each figure and table you are considering including in the poster. Does it relate to your stated objectives? Does it directly support the results you have chosen to emphasize? If not, then don't use it. Irrelevant, repetitive graphs and tables will overwhelm readers just as readily as dense passages of text. Scrutinize the visual elements you do plan to use, and streamline and simplify them wherever possible. Never simply repeat in words what is already shown visually.

Figure legends and titles of tables can often be eliminated if the content is incorporated into your accompanying text. Similarly, numbering of your figures and tables should not be necessary; just place these materials at appropriate points in the logical flow of your story. Make all graphics as self-explanatory as possible.

Keep cited references to a minimum, perhaps just two or three, if you can manage it. Viewers rarely read this section.

■ Make contents readable from a distance.

Think on a large scale. Your title should be readable from a distance of 15–20 feet, so choose a large font size (e.g., 72–84 point). The text of your poster, as well as numbers and words in figures and graphs, should be readable from about 6 feet away; here, 24-point type is commonly used, with 36-point type for section headings. For literature cited and acknowledgments, you can use 18-point type to save space for other sections.

Double-space your text and align type on the left but not on the right, for easier reading. For text, many people prefer a serif font such as Times New Roman for readability; others prefer sans serif fonts such as Arial or Verdana. Consider the restrained use of boldface, color, or larger type sizes to announce different sections of the paper or to emphasize key points.

■ Minimize distracting elements. Be consistent in your presentation.

For consistency, stay with a single font throughout the text of your poster. (Some people adopt a different one for the title.) Similarly, use the same font size for the headings of different sections, and use parallel grammatical form for headings as well as within lists of related points.

If you choose to use a colored background, use just a single color to tie all elements together. Some people frame various sections of the poster with borders of a different color to set them off from the background. However, keep in mind that overuse of different colors, as well as excessive framing, can be distracting. Avoid using any design gimmick, including clip art, simply for an unusual effect.

In summary, strive for a coherent, attractive visual product that does not overpower or distract from the real substance of your poster. Remember that your main task is to communicate your findings.

■ Solicit feedback on a rough draft.

Allow time to prepare a preliminary version of your poster—ideally a life-sized model of exactly what you intend to present. Ask your instructor, classmates, or other willing critics to comment on your poster's readability, clarity, coherence, layout, and overall impact. Also solicit their feedback about the research itself.

■ Prepare for informal discussion and questions.

Visitors to your poster will probably expect you to speak informally about your work, giving them a brief (several minutes) tour of the poster. Prepare notes in advance, but don't refer to your notes when speaking, and don't simply read back the poster out loud. Some viewers will also have questions; try to anticipate these and rehearse brief replies. You may wish to prepare supplementary handouts giving more details about your methods, related findings not covered by the poster, or other information. Be sure to include your name, academic address, and e-mail address. Authors of conference posters sometimes bring personal business cards or reprints of papers they have written on similar topics. One of the chief attractions of scientific conferences in general and poster sessions in particular is that they help you meet other people whose work may be relevant to your own.

■ Checklist for poster presentations

- Have you followed all guidelines about poster size, content, and other details?
- Does your poster address the background and expectations of your audience?
- Do you use a logical organization, following the plan of a scientific paper?
- Do you clearly state your objectives or hypothesis?
- Have you reduced your materials and methods section to the bare essentials?
- Do you present only the most important results?
- Have you included only those figures and tables that directly support these results?
- Have you briefly interpreted your findings in the context of established literature?
- Do you keep cited references to a minimum?
- Are all contents readable from a distance?
- Is your layout easy to follow? Are you consistent and restrained in the use of color, boldface, and other design features?
- Does your poster design emphasize and enhance the contents rather than distract from them?

- Have you obtained feedback from others on a rough draft?
- Are you prepared for informal discussion and questions?

SAMPLE POSTER PRESENTATION

Following is a poster presentation of the student research project in Chapter 4 (pp. 102–113). Notice how the poster includes more details than the PowerPoint presentation of the same work (pp. 219–224) but is much less detailed than the original paper. Also, the title of the paper has been rephrased to attract viewers to the poster by presenting a strong statement of the student's results.

Gibberellins Increase Germination of

Kristin A. VanderPloeg, Dept.

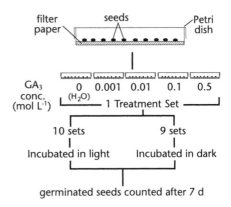

Gibberellic acid (GA_3)

Introduction

- In many plants, seed germination depends on light, which stimulates the synthesis of gibberellins (GAs) needed for embryonic development. Seeds of some light-requiring species will germinate in darkness if provided with exogenous GAs.
- The *gal-2* and *gal-3* mutants of *Arabidopsis thaliana* cannot synthesize active GAs at all; hence, they require exogenous GAs for germination in both light and darkness. Working with the *gal-2* mutant, Derkx and Karssen (1993a) found that, in the presence of light, a lower concentration of GA is needed to produce germination than in darkness. This finding suggests that *gal-2* seeds are more sensitive to GAs in light than in darkness.

Objectives

I investigated the sensitivity of seeds of the *gal-3* mutant of *A. thaliana* to varying concentrations of gibberellic acid (GA_3) in both light and darkness.

Materials and Methods

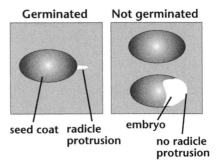

Criterion for germination

Germinated	Not germinated

seed coat   radicle protrusion   embryo   no radicle protrusion

Arabidopsis thaliana in Light and Darkness
of Biology, Colgate University

Results and Discussion

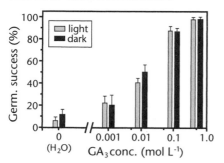

- Germination success increased significantly with increasing GA_3 concentration in both light and darkness. Similar findings have been reported for wild-type and *gal-2* mutants of *A. thaliana* (Derkx and Karssen 1993a,1993b).

Two-way ANOVA: GA_3 concentration and light/dark conditions

Source of variation	F
GA_3 conc.	105.96***
Light/Dark	0.41 ^{ns}
GA_3 conc. x Light/Dark	0.48 ^{ns}

*** = $P < 0.001$; ns = nonsignificant ($P \geq 0.05$)

Tukey tests showed that each successive increase in GA_3 concentration produced a significantly greater germination success in both light- and dark-treated seeds ($P < 0.05$).

- At all GA_3 concentrations, germination success did not differ in light *vs.* darkness. These results conflict with those of Derkx and Karssen (1993a), who found that seeds of the *gal-2* mutant had a higher germination success in light than in darkness at the same GA concentrations used here.

Possible explanations for this discrepancy:

a) Differences in light sensitivity between mutants

b) Brief exposure to fluorescent light during handling of all seeds

Literature Cited

Derkx MPM, Karssen CM. 1993a. Effects of light and temperature on seed dormancy and gibberellin-stimulated germination in *Arabidopsis thaliana*: studies with gibberellin-deficient and -insensitive mutants. Physiol Plant. 89(2):360-368.

Derkx MPM, Karssen CM. 1993b. Variability in light-, gibberellin- and nitrate requirement of *Arabidopsis thaliana* seeds due to harvest time and conditions of dry storage. J Plant Physiol. 141(5):574-582.

Acknowledgments

I thank C. LaFave for advice during the research.

RESEARCH PROPOSALS

Research proposals have many traits in common with research papers (see Chapter 4): both introduce a scientific question or hypothesis, both put a specific study in the broader context of existing research, and both are organized in ways that reflect the logic of the scientific method. Of course, a research proposal is missing a Results section, along with the author's analysis and interpretation of those results. Instead, it seeks to propose and justify the author's research *plans*—perhaps for a senior project, a summer internship, a graduate thesis, or, in the case of professional biologists, new or continuing work in a specialized field.

There is no one standard format for a research proposal. Professional scientists seeking funding from the National Science Foundation or other agencies must adhere to the precise guidelines laid down by those organizations; similarly, student authors should follow the guidelines applicable in their own situations. All research proposals involve similar basic challenges and constraints. If you are presented with the opportunity to do research, the following guidelines should help you in preparing your own proposal.

■ Create a specific and informative title.

A focused title suggests a well-thought-out project. Titles such as "A Study of Stream Ecology" or "Nutrient Uptake in Seaweeds" convey little sense of direction or purpose; compare them with more specific versions: "The Effect of Open and Closed Canopy on the Diversity of Macroinvertebrates in Morrisville Creek"; "Ammonia Uptake by *Ulva curvata* Growing on Three Different Substrates." You may need to modify your title as you proceed with the research, but a good working title is essential to a well-received proposal.

■ Start with an overview of the general scientific issue.

Before proposing your own study, you need to introduce the broader topic to which your work relates. Most research proposals begin with an Introduction (or Background to the Study) section, which presents a theoretical context for the proposed investigation by summarizing and integrating the work of others in the field. Doing a good job on this part involves a thorough literature search and careful, selective use of primary sources, which are cited in your text and then listed later in a Literature Cited section. Your introductory discussion should also reflect a sensitivity to the scientific background of your readers. Typically, you'll be submitting your proposal to other scientists; however, if it will be judged by nonscientific readers (for example, a multidisciplinary committee), then keep specialized vocabulary to a minimum and use a less technical style.

■ State the aims of your study, and provide a clear rationale.

After providing background material and highlighting major findings, you can then turn to important gaps in our knowledge. What conflicts exist in the literature? What questions remain? Relate such issues to your *own* proposed study and its specific aims. What questions will you investigate? How will your work enlarge, clarify, or complement existing knowledge of the subject? Discussion of your own objectives can come at the end of the Introduction or Background section or in a subsequent section.

Remember that you are trying to convince people to believe in your project strongly enough so that they will give you money to carry it out. Your proposal should present a compelling argument for the potential value of your work.

■ Summarize your methods.

Next explain specifically how the research will be conducted, what kinds of data you will collect, and how those data will be analyzed. The level of detail, and thus the amount of space devoted to the Methods section, will depend on the type of proposal you are submitting and the audience for whom it is intended. Remember that even the most promising research question will fall flat if you can't devise a practical and scientifically valid way to investigate it. Think through your methods completely, and become thoroughly familiar with the procedures used in studies similar to yours. If you want to convince your readers to commit their time and/or money to your project, then it must sound feasible.

■ Provide a budget, if necessary.

Here, too, follow whatever specific guidelines apply to your situation. If an estimate of projected expenses is requested, prepare this with painstaking care. You will need to identify potential suppliers, compare prices, and compile a detailed list. Don't be vague: "Collecting supplies— $100"; instead, give a breakdown of exactly what you'll need and how much each item will cost: "1 dip net, $38.50; 2 replacement bags @ $8.98 each; 1 D-frame aquatic net, $39.95. . . ." This task may take much time and effort; however, it will ensure that you consider your methodology very carefully. Moreover, your proposal will more likely be accepted if you show that you can do what you propose within the budget allotted to you.

■ Follow application instructions carefully, and submit your proposal on time.

This advice may seem obvious, yet many applicants run the risk of having their proposals rejected out of hand because they fail to include all required materials. Pay meticulous attention to such requirements as page length, format, margins, spacing, and the number of copies to be submitted. Check to see if your résumé and/or letters of recommendation are also required. You may also need to prepare a brief cover letter to accompany your proposal. Finally, be prepared to meet the application deadline; failure to do so suggests immediately that you are incapable of adequate planning and a professional approach.

SAMPLE RESEARCH PROPOSAL

Following is a sample research proposal for an undergraduate summer research fellowship.

Field Measurements of
Photosynthesis and Transpiration
Rates in Dwarf Snapdragon
(*Chaenorrhinum minus* Lange):
An Investigation of Water Stress
Adaptations

Tara Gupta

Application for Summer Research
Fellowship,
Colgate University
March 12, 200-

Compose a specific and informative title. Include your name and other relevant information on a title page.

Water Stress Adaptations 2

INTRODUCTION

Introduce the
scientific issue
and give back-
ground infor-
mation. Cite
relevant studies
by others;
CSE citation-
sequence (num-
ber) system is
illustrated here.
Adjust vocabu-
lary to take into
account the bi-
ological back-
ground of your
readers.

Dwarf snapdragon (*Chaenorrhinum
minus*) is a weedy pioneer plant found
growing in central New York during spring
and summer. Interestingly, the distribution of this species has been limited almost exclusively to the cinder ballast of
railroad tracks [1], a harsh environment
characterized by intense sunlight and
poor soil water retention. Given such environmental conditions, one would expect
C. minus to exhibit anatomical features
similar to those of xeromorphic plants
(species adapted to arid habitats).

However, this is not the case.
T. Gupta and R. Arnold (unpublished,
2004) have found that the leaves and
stems of *C. minus* are not covered by a
thick, waxy cuticle but rather by a thin
cuticle that is less effective in in-
hibiting water loss through diffusion.
The root system is not long and thick,
capable of reaching deeper, moister
soils; instead, it is thin and diffuse,
permeating only the topmost (and driest)
soil horizon. Moreover, in contrast to
many xeromorphic plants, the stomata
(pores regulating gas exchange) are at
the leaf surface, not found in sunken
cavities in the epidermis that retard
water loss from transpiration.

Despite a lack of these morphological adaptations to water stress, *C. minus*
continues to grow and reproduce when

morning dew has been its only source of
water for up to five weeks (R Arnold,
personal communication). Such growth involves fixation of carbon by photosynthesis and requires that the stomata be open
to admit sufficient carbon dioxide. Given
the dry, sunny environment, the time required for adequate carbon fixation must
also mean a significant loss of water
through transpiration as open stomata exchange carbon dioxide with water. How
does *C. minus* balance the need for carbon
with the need to conserve water?

AIMS OF THE PROPOSED STUDY

State aims and scope of proposed study.

The above observations have led me
to explore the extent to which *C. minus*
is able to photosynthesize under conditions of low water availability. It is my
hypothesis that *C. minus* adapts to these
conditions by photosynthesizing in the
early morning and late afternoon, when
leaf and air temperatures are lower and
transpirational water loss is reduced. I
predict that its photosynthetic rate may
be very low during the middle of the day,
perhaps even zero on hot, sunny afternoons. Similar diurnal changes in photosynthetic rate in response to midday
water deficits have been described in
crop plants [2,3]. There is only one compara-
ble study [4] on noncrop species in their
natural habitats.

Thus, the research proposed here
should help explain the apparent paradox

Water Stress Adaptations 4

of an organism that thrives in water-stressed conditions despite a lack of morphological adaptations. This summer's work will also serve as a basis for controlled experiments in a plant growth chamber on the individual effects of temperature, light intensity, soil water availability, and other environmental factors on photosynthesis and transpiration rates. These experiments are planned for the coming fall semester.

METHODS

Briefly describe your methodology.

Simultaneous measurements of photosynthesis and transpiration rates will indicate the balance *C. minus* has achieved in acquiring the energy it needs while retaining the water available to it. These measurements will be taken daily at field sites in the Hamilton, NY, area, using an LI-6220 portable photosynthesis system (LICOR, Inc., Lincoln, NE). Basic methodology and use of correction factors will be similar to that described in related studies [5-7]. Data will be collected at regular intervals throughout the daylight hours and will be related to measurements of ambient air temperature, leaf temperature, relative humidity, light intensity, wind velocity, and cloud cover.

BUDGET

If a budget is required, be as specific as possible.

1 kg soda lime $53.90
   (for absorption of CO_2 in photosynthesis analyzer

Water Stress Adaptations 5

1 kg anhydrous magnesium perchlorate (used as desiccant for photosynthesis analyzer)	$274.40
Shipping of chemicals (estimate)	$12
Estimated 500 miles travel to field sites in own car @ 40.5¢/ mile	$202.50
CO_2 cylinder, 80 days rental (for calibration of photosynthesis analyzer)	$100
Total request	$642.80

REFERENCES

1. Widrlechner MP. Historical and phenological observations of the spread of *Chaenorrhinum minus* across North America. Can J Bot. 1983;61(1):179-187.

2. Manhas JG, Sukumaran NP. Diurnal changes in net photosynthetic rate in potato in two environments. Potato Res. 1988;3(2):375-378.

3. Yordanov I, Tsonev T, Velikova V, Georgieva K, Ivanov P, Tsenov N, Petrova T. Changes in CO_2 assimilation, transpiration and stomatal resistance in different wheat cultivars experiencing drought under field conditions. Bulg J Plant Physiol. 2001;27(3-4):20-33.

4. Chaves MM, Pereira JS, Maroco J, Rodrigues ML, Ricardo CPP, Osório ML, Carvalho I, Faria T, Pinheiro C. How plants cope with water stress in the

Include all published works cited. Numbers correspond to the order in which sources were first mentioned in the text. Author's last name is followed by initials, then paper title, journal, publication date, volume, issue, and page numbers.

Water Stress Adaptations 6

   field: photosynthesis and growth.
   Ann Bot. 2002;89(Jun):907-916.
5. Jarvis A, Davies W. The coupled response of stomatal conductance to
   photosynthesis and transpiration.
   J Exp Bot. 1998;49(Mar):399-406.
6. Kallarackal J, Milburn JA, Baker DA.
   Water relations of the banana.
   III. Effects of controlled water
   stress on water potential, transpiration, photosynthesis and leaf
   growth. Aust J Plant Physiol.
   1990;17(1):79-90.
7. Idso SB, Allen SG, Kimball BA,
   Chouhury BJ. Problems with porome-
   try: measuring net photosynthesis by
   leaf chamber techniques. Agron J.
   1989;81(3):475-479.

LETTERS OF APPLICATION

Letters of application, along with cover letters accompanying résumés, are never mere formalities. Prospective employers or graduate school admissions committees do read such letters, often before looking at any other material about you. If you want to create a good first impression, then spend as much time as possible on this stage of the application process. Draft *each* cover letter carefully, revise it thoughtfully, and proofread it meticulously. Use the letter to grab the reader's attention, identify yourself as a candidate, highlight your best qualities, and demonstrate a sincere interest in the position for which you are applying. A compelling cover letter will distinguish you from the pool of other applicants; it also may offset, in the reader's mind, any deficiencies in your qualifications. The following guidelines should help you.

■ Use a standard business letter format.

One variation of this format is illustrated in the sample letter on page 245. Using a computer, type your return address (omitting your name), followed by the date, in the upper right-hand corner; then space down three or four lines, and at the left margin type the name and address of the person to whom you are sending the letter. Double-space; then type the salutation (Dear Mr. ———, Ms. ———, Dr. ———, or other appropriate greeting) followed by a colon. Double-space again and begin the body of your letter, single-spaced, with double-spacing between paragraphs (optional). Indent each paragraph five spaces. After the last paragraph, double-space and type a standard closing (Sincerely, Yours truly, or the like), followed by a comma; then skip four lines and type your name. The closing and your name are placed at the right-hand side of the letter, in line with the return address. In the space between the closing and your typed name, add your written signature, using your full name.

A variation is to type all components of the letter (return address, date, closing, your name, and so on) flush with the left margin. If you choose to prepare your cover letter this way, you do not indent paragraphs (but be sure to double-space between them).

■ Print the letter on appropriate stationery.

Do not write out the letter in longhand. Use good-quality, unlined paper—8½ × 11 inches is a standard size; never use informal personal stationery. Print on only one side of the page. Remember that your cover letter should convey a sense of seriousness and professionalism.

■ Be straightforward and concise.

Restrict your letter to a single page, if possible. Your reader may be screening a large number of applicants and may have neither the time nor the patience to deal with a long treatise about you and your many qualifications. Remember the purposes of a cover letter: to identify yourself as an interested candidate and to highlight your strengths. There is no need to paraphrase your entire résumé.

■ Use specific language and details.

Experienced readers can quickly spot the form-letter approach used by some applicants to save time and effort. This tactic involves composing a standard, general cover letter and sending it out (often along with a standard, general résumé) to a large number of addresses. Consider the following example:

```
Dear Department Chair:

    I am a senior biology major at Shoreham Col-
lege. I find your university very interesting
and would like to apply to the Ph.D. program.
Please also consider me for a teaching assis-
tantship. Enclosed is my résumé.
    Thank you.

                              Yours truly,
```

Form-written cover letters often do you more harm than good. If you want to stand out from other applicants and convince your reader that you are sincerely interested in this *particular* job or opening, then you must individualize each letter. What is it about this specific university, program, business, or agency that draws you to it? What special aspects of your background or training suit you for this opening? What makes you a unique applicant for the position? Do not speak in vague generalities: shape your letter to fit the context, and support your points with specific details. Strive for a positive, confident tone without sounding boastful. Of course, you must write honestly and sincerely; do not say anything that you may have to back down from later.

Wherever possible, send your letter to the person who is *directly* involved in the hiring or acceptance process; otherwise, your letter may get sidetracked. If you do not have this person's name, call or email the business or university to obtain it. Be sure to check the spelling of the person's name to whom you are sending your letter.

The following sample is a letter of application for a research assistant position. Note how the writer has tailored her letter to a particular reader and to the specifics of the job situation.

SAMPLE COVER LETTER

Box 7713
Gorton College
Roslyn Fields, New York 13655
April 17, 2005

Professor Patrick Hayes
Department of Biological Sciences
Johnson State College
Chandlerville, Maryland 21323

Dear Professor Hayes:

I am a junior at Gorton College, where I am majoring in biology. I am interested in animal behavior, particularly field studies of insects, and hope to eventually do research of my own in this area as a graduate student. Professor Ruth Swarthout, my academic advisor, first introduced me to your papers on dragonfly territoriality and suggested that I contact you. She told me that you sometimes hire undergraduates during the summer to assist with field work and data collection. I am writing to ask if I might be considered for such a position.

My coursework at Gorton College has given me a good background in animal behavior and ecology, as well as laboratory and field experiences. For example, in Biology 255 (Entomology), I learned how to work with many different taxonomic keys, collecting and mounting over 100 species for a personal insect collection. (Since then, I have added over 30 additional specimens on my own.) In addition, I worked with two other students on a lab project investigating sex differences in hissing cockroaches (*Gromphadorhina portentosa*). In Biology 301 (Aquatic Insects), I learned basic freshwater sampling techniques, along with quantitative methods such as ANOVA and factor analysis, during a class study of species diversity at local streams. I am currently taking Professor Swarthout's advanced seminar, Biology 401 (Topics in Evolutionary Ecology), for which I am writing a 20-page term paper on the evolution of eusocial behavior in termites.

Last summer I worked for Dr. Andrea Rider, at Gorton College, assisting with her research on predatory behavior in waterscorpions (Nepidae). I made routine field collections, maintained captive insects in the lab, and helped gather and analyze data on feeding preferences. During my last month at work, I performed most of these tasks with little supervision. This job showed me that behavioral research requires long periods of concentration, attention to details, and the ability to be flexible when things don't always go as planned. For these reasons, I think I could be a reliable and dedicated assistant in your research program.

My parents live just outside Chandlerville, about six miles from Johnson State College. At the end of this semester, I will be returning home to Maryland for the summer and would be available for full-time employment through August 25. I will have the use of a car if independent field work is required. I will be at Gorton College until May 30, after which I can be reached at home (25 Hamilton Lane, Chandlerville, MD 21323; home phone, 842-343-3576). I can also be reached via email (esjarvis@mail.gorton.edu).

Enclosed is a copy of my résumé. Thank you for considering me as a summer research assistant. I look forward to hearing from you.

Sincerely,

Emma Jarvis

Emma Jarvis

RÉSUMÉS

Composing an effective résumé may be one of the most important writing projects you'll face during your academic or professional career. If carefully crafted, your résumé may be instrumental in getting you that first interview and a job or acceptance in a graduate program. If poorly prepared, your résumé may generate little response from its readers or even elicit a strongly negative reaction, damaging your chances to be considered further.

Your basic task in writing a résumé is to promote yourself as a candidate by highlighting your most important achievements and qualifications. A résumé usually begins with your name, address, and phone number, followed by such information as your career objectives, educational background, employment history, research and teaching experience, awards received, and other relevant interests, skills, or activities. You are not required to give personal information such as sex, race, religion, marital status, or age. There is no one right way to put all this material together; however, the following suggestions will apply to the preparation of most résumés.

■ Use a professional-looking and inviting format.

Your résumé should be typed on a computer and printed on good quality paper to match your cover letter. Organize information using headings (and, if necessary, subheadings); the sample résumé on page 249 illustrates a common format. Notice the use of single- and double-spacing to tie together related material and to separate successive sections. Use capitalization, boldface type, underlining, indentation, or other devices to organize your text, but be sure to be consistent. A chronological approach is often used to list related information, such as jobs held or degrees earned, starting with the most recent items and working backward. However, do not feel bound to this convention. If, for example, work you did five years ago is most relevant to the position you want, then list this employment first, followed by other work of less importance.

■ Modify your résumé to fit each application.

Most of the material in your résumé will be relatively fixed—for example, your educational history, much of your work experience, and so on. However, think carefully about each position for which you are applying. Some skills and experiences may be more relevant in some situations than in others. Look again at the sample résumé. The fact that the applicant is fluent in Japanese probably has little bearing on her chances for a research assistant job in Maryland; however, if she were applying to a student exchange program in Tokyo, she would do well to expand a description of this skill. In addition to the camp counselor work already listed, she might add that she studied Japanese for three years in high school and two years at college, and also spent three weeks in Japan one summer. Similarly, the cover letter accompanying her résumé should highlight these and other pertinent parts of her background (see p. 245).

Thus, to tailor your résumé to fit each position, you need to think creatively about yourself. It's easy to overlook (or to dismiss as unimportant) such experiences as unpaid work at a school or hospital, membership in special-interest clubs, athletic or musical abilities, evening or weekend classes, relevant hobbies, and so on that might be of interest. Put yourself in the position of your prospective employer, school, or program, and try to imagine the "package" of qualities you might look for in a candidate. Your résumé should present your particular version as convincingly as possible.

■ Be concise and selective.

Remember that you'll probably be competing with many other applicants; consequently, your readers may take only a quick glance at your credentials. In any event, most people will be put off by a lengthy résumé unless every item in it is pertinent and noteworthy. Never pad your résumé with unnecessary details or other material just to fill up space; experienced readers will quickly suspect that you are trying to look better on paper than you really are. A good résumé is straightforward, succinct, and short— typically a single page. On the other hand, a curriculum vitae, which contains more detailed information about your coursework, field experiences, lab skills, and other academic qualifications, may be two pages or longer. You will probably want to prepare a curriculum vitae rather than a résumé if you are applying to graduate school or seeking fellowships or other academic opportunities.

■ Update your résumé frequently.

Add new accomplishments, skills, and experiences as you acquire them; at the same time, modify or delete outdated material.

■ Edit and proofread meticulously.

Even though you will be working mostly with lists and sentence fragments, choose your words carefully. Use parallel grammatical forms for lists of similar items. Remember that even a single misspelled word or typographical error will detract from the professional appearance you want to convey. Give your résumé to other people to proofread (and critique). Try reading it aloud to spot errors.

■ **Supply references.**

This may be done in several ways. Your résumé can include the names, addresses, and phone numbers of several people who may be contacted for letters of recommendation. Another option is to give this information in your cover letter. A third method is simply to say, "References available on request." You will then be contacted for more information if your prospective employer or program is interested in your application. Be sure to check the application instructions or employment advertisement to see if there are special guidelines about how or when references should be supplied.

Also, of course, you'll need to obtain the permission of each person whom you wish to list as a reference. Remember that such people should be familiar with at least some of your abilities and strengths; good choices include former employers and instructors in whose courses you distinguished yourself. Give these people plenty of advance notice—at least a few weeks—and make things easier for them by supplying a copy of your résumé, information about the position, and an addressed, stamped envelope to use in mailing their recommendation. Be sure to inform them of the application deadline. It's also a courteous gesture to follow up with a brief note of thanks, whether or not your application was successful.

The following sample student résumé should give you some idea of what a well-crafted and well-prepared résumé should look like. See also the sample curriculum vitae on pages 250–252.

SAMPLE RÉSUMÉ

EMMA JARVIS
25 Hamilton Lane
Chandlerville, MD 20723
842-567-0328
ecjarvis@mail.gorton.edu

CAREER GOALS	Graduate study, followed by a university teaching and research position in animal behavior. *Special interest in insect behavior.*
EDUCATION	Gorton College, Roslyn Fields, NY. Bachelor of Science expected May 2006. *Major in biology.*
ADDITIONAL STUDY	University of Wyoming Field Station, Clemow Lake, WY. Summer 2001. *Five-week field and laboratory experience in biology for high school students.*
RESEARCH EXPERIENCE	Research assistant in animal behavior for Dr. Andrea Rider, Biology Department, Gorton College, summer 2004. *Collected and maintained captive insects; assisted with data collection and analysis.*
TEACHING EXPERIENCE	Undergraduate teaching assistant, Zoology 101, Gorton College, fall 2004. *Took major responsibility for setting up labs; also conducted tutoring sessions and assisted with grading.*
ADDITIONAL WORK EXPERIENCE	Office assistant, Biology Department, Gorton College, spring 2004. *Assisted with typing, filing, and photocopying.* Camp counselor, Summer Nature Program, Warbler Lane Nature Preserve, Chandlerville, MD, summer 2003. *Supervised group of 10–12 elementary school children attending 6-week day camp; taught basic nature appreciation and nature crafts.* Volunteer, Bluffton Animal Shelter, Bluffton, MD, summer 2002. *Assisted with animal care and public awareness programs.*
HONORS	Huerta Prize for Scientific Writing (best lab report by first-year student), 2003. Dean's Award for Academic Excellence (cumulative GPA of 3.5 or higher), 2002–2004.
SPECIAL SKILLS AND INTERESTS	Certified by AHA for CPR and First Aid instruction, May 2004–present. Competent in use of Microsoft Word, JMP, Adobe Photoshop. Fluent in Japanese. Participant (actor, singer) in community and college theater productions, 2000–present.
REFERENCES	AVAILABLE ON REQUEST.

SAMPLE CURRICULUM VITAE

STEPHEN R. PORTER
arp@center.silsbyc.edu

31 Cold Spring Road
St. Andrews, NY 13340
315-824-6162

Silsby College
Ellicott, CT 06269
861-486-1000

OBJECTIVE

To obtain an internship or research assistant position in the health sciences.

EDUCATION

Silsby College, Ellicott, Connecticut, 2005. Bachelor of Arts with major in Biology.
GPA: 3.8
University of New South Wales, Sydney, Australia, fall semester, 2004.

HONORS

President's Award for Academic Excellence: 2002–05.
Tri-Beta National Biological Honors Society: inducted October 2004.

COURSEWORK

Introductory Biology, Ecology, Evolution and Taxonomy of Vascular Plants, Genetics, Cell
Biology, Animal Physiology, Invertebrate Zoology, Medical Entomology, Scientific Writing,
Independent Research in Biology, Biostatistics, General Chemistry I, II, Organic Chem-
istry I, II, Physics I.

RESEARCH EXPERIENCE

Independent Research Project, Silsby College. Performed semester-long study for
course credit on the effects of temporary food deprivation on eating behavior in mice.
Spring 2004.
Summer Research Assistantship, Clinical Research Department, Beaufort Community
Hospital, Beaufort, CT. Assisted with research on Lyme disease. Responsibilities included
running Western Blot, ELISA, and PCR tests. Summer 2003.

PRESENTATIONS

Porter SR. Effects of temporary food deprivation and eating behavior in mice (*Mus muscu-
lus*). Honors presentation, Silsby College. April 17, 2005.
Porter SR, Arnold J. Student opportunities in emergency medicine. Invited presentation,
Northeastern Society of Volunteer Emergency Medical Technicians, 15th Annual Meeting, New York, NY. June 2, 2004.

LABORATORY EXPERIENCE

Equipment Skills
Trained in the use of microscope, pH meter, centrifuge, autoclave, and spectrophotome-
ter. Experienced in techniques of chromatography (thin layer, column, GC), nuclear
magnetic resonance, PCR, and electrophoresis.

Animal Care
Trained in the maintenance and breeding of laboratory mice and golden hamsters.

HEALTHCARE EXPERIENCE

Health Sciences Intern, Upstate Family Practice, Ellicott, CT. Shadowed family physician and gained exposure to the fields of pediatrics, gerontology, Ob-Gyn, and general medicine. Spring 2005.

Emergency Medical Technician, Ellicott Volunteer Ambulance Corps, Ellicott, CT. Helped provide emergency medical care to local residents. Fall 2002–Spring 2005.

Health Sciences Intern, Long Island Dermatology Center, St. Andrews, NY. Observed minor surgery and clinical patient care; assisted with some procedures. Summer 2004.

Health Sciences Intern, New York Center for Eating Disorders, Caligny, NY. Shadowed physicians and therapists; assisted in obtaining case histories and enrolling patients in studies. Summer 2002.

Volunteer, Morrisville Senior Residential Community, Morrisville, CT. Assisted nursing staff at assisted living residence. Summer 2000.

TEACHING EXPERIENCE

Student Teaching Assistant, Invertebrate Zoology. Assisted with laboratories and field trips; helped devise lab quizzes; conducted review sessions for students. Fall 2004.

Peer Tutor (Sciences), Interdisciplinary Writing Center, Silsby College. Instructed first-year students on the preparation of lab reports for introductory biology and chemistry courses. Spring 2004

WORK EXPERIENCE

Resident Advisor, Silsby College. Served as mentor to younger residential students; enforced university alcohol and drug regulations; devised weekly educational and social programs. 2004–05.

Assistant Coach, Southeast Soccer Academy, Garvies Point, SC. Helped coach teenagers at 8-week overnight soccer camp. Summer 2001.

COMPUTER SKILLS

SAS, SigmaPlot, Microsoft Word, Excel, PowerPoint.

EXTRACURRICULAR ACTIVITIES

Silsby Community Outreach Program. Participated in fundraising events to aid local programs for the elderly. Fall 2003.

Big Brother Program, Silsby College. Served as a role model to an elementary school student two afternoons per week. 2002–03.

Biology Club, Silsby College, 2002–05. Treasurer, Spring 2003.

Varsity Soccer Team, Silsby College, 2001–03.

REFERENCES

Dr. Niamh Corbet
Biology Department
Silsby College
Ellicott, CT 06269

Dr. Elizabeth Ketchum
Upstate Family Practice
Ellicott, CT 06269

Dr. Alice Johnson
Long Island Dermatology Center
St. Andrews, NY 13340

Additional Readings

General Writing Handbooks

Every writer should have at least one general writing guide. The following handbooks cover the essentials of grammar, punctuation, and word choice as well as the crafting of sentences, paragraphs, and whole essays. All are organized as self-help manuals presenting easy-to-follow guidelines applicable to a wide variety of writing situations.

Hacker D. 2006. The Bedford handbook. 7th ed. Boston: Bedford/ St. Martin's. 874 p.

Hacker D. 2003. A writer's reference. 5th ed. Boston: Bedford/ St. Martin's. 466 p.

A concise and easy-to-use handbook that offers thorough coverage of a range of topics, including composition and revision, grammar and mechanics, and research and documentation.

Other classic, useful books on the practice of writing:

Baker S. 1998. The practical stylist. 8th ed. New York: Longman. 268 p.

Hairston MC. 1998. Successful writing. 4th ed. New York: Norton. 246 p.

Strunk W, White EB. 2000. The elements of style. 4th ed. Needham Heights (MA): Allyn & Bacon. 105 p.

Guides to the Research Process

Hacker D. 2006. Research and documentation in the electronic age. 4th ed. Boston: Bedford/St. Martin's. 269 p.

This guide provides an introduction to the Internet with detailed advice for evaluating electronic sources and guidelines for documenting print and online sources. The online version, available at <http://www .bedfordstmartins.com/hacker/resdoc>, includes the complete text of the booklet and links to all the Internet sites it lists.

Harnack A, Kleppinger E. 2003. Online! a reference guide to using internet sources. Boston: Bedford/St. Martin's. 260 p.

In addition to providing models for citing and documenting Internet sources, this guide offers extensive advice on navigating the Internet, conducting Web searches, and evaluating sources. The companion Web site, located at <http://www.bedfordstmartins.com/online>, provides models for citing and documenting electronic sources, as well as links to other Web sites for online citation styles.

Palmquist M. 2006. The Bedford researcher. 2nd ed. Boston: Bedford/St. Martin's. 383 p.

Guides to Scientific Writing

The references that follow are particularly useful for more experienced scientific writers and professional biologists and for undergraduate and graduate students preparing a paper for a technical journal.

The Chicago manual of style. 2003. 15th ed. Chicago: University of Chicago Press. 984 p.

A widely used reference for writers in many academic fields. Covers a wide range of topics, including documentation methods, tables and figures, use of quotations, and many aspects of writing mechanics and style. Useful as a supplement to guides specifically devoted to scientific writing.

Council of Science Editors Style Manual Committee. 2006. Scientific style and format: the CSE manual for authors, editors, and publishers. 7th ed. New York: Rockefeller University Press.

This standard and indispensable reference for biologists contains detailed instructions for preparing scientific manuscripts for publication. Topics include punctuation, grammar, prose style, manuscript format, tables and figures, chemical notation, cells and genes, symbols and abbreviations, documentation methods including those for the Internet, scientific nomenclature, and many others. The manual can be found in the reference section of your library.
For more information about this reference, see the CSE Web site at <http://www.councilscienceeditors.org/>

Iverson C, Flanagin A, Fontanarosa PB, et al., editors. 1997. American Medical Association manual of style. 9th ed. Baltimore: Lippincott Williams & Wilkins. 660 p.

American Psychological Association. 2001. Publication manual of the American Psychological Association. 5th ed. Washington: APA. 439 p.

Guides to Scientific Literature

Alpi KM. 2001. Science & technology resources on the Internet. What you see is what you get: science images on the Web. Issues Sci Technol Librariansh [Internet]. Available from <http://www.istl .org/01-summer/internet.html>

Mathews BS. 2004. Gray literature: resources for locating unpublished research. Coll Res Libr News. 3:125–8.

Boorkman JA, Huber JT, Roper FW, editors. 2004. Introduction to reference sources in the health sciences. 4th ed. New York: Neal-Schuman Publishers. 389 p.

Statistics

The following texts are useful for beginners:

Glover T, Mitchell M. 2001. An introduction to biostatistics. New York: McGraw-Hill. 432 p.

Hampton RE. 2003. Introductory biological statistics. Long Grove (IL): Waveland Press. 233 p.

The texts below, written for advanced readers, give a comprehensive, thorough treatment of many statistical topics:

Sokal RR, Rohlf RJ. 1994. Biometry: the principles and practice of statistics in biological research. 3rd ed. New York: Freeman. 880 p.

Zar JH. 2005. Biostatistical analysis. 5th ed. Englewood Cliffs (NJ): Prentice-Hall. 960 p.

For users of JMP or JMP-IN:

Sall J, Lehman A, Creighton L. 2001. JMP Start Statistics: a guide to statistical and data analysis. Pacific Grove (CA): Duxbury/Thomson Learning. 491 p.

Figures and Tables

Cleveland WS. 1985. The elements of graphing data. AT&T Bell Laboratories. New York: Chapman & Hall. 297 p.

Oral Presentations

Battalio J. Preparing presentation slides: a tutorial. Available from <http://bcs.bedfordstmartins.com/techcomm/content/cat_030/ preparingpresentationslides/2g.html>

Literature Cited

Bem SL. 1981. Gender schema theory: a cognitive account of sex typing. Psychol Rev. 88(4):354–364.

Burger J. 1974. Breeding adaptations of Franklin's gull (*Larus pipixcan*) to a marsh habitat. Anim Behav. 22(3):521–567.

Dawkins R. 1976. The selfish gene. New York: Oxford University Press. 368 p.

Eiseley L. 1958. Darwin's century: evolution and the men who discovered it. New York: Doubleday. 449 p.

Ginsberg HJ, Braud WG, Taylor RD. 1974. Inhibition of distress vocalizations in the open field as a function of heightened fear or arousal in domestic fowl. Anim Behav. 22(3):745–974.

Hepler PK, Wayne RO. 1985. Calcium and plant development. Ann Rev Plant Physiol. 36:397–439.

Kohmoto K, Kahn ID, Renbutsu Y, Taniguchi T, Nishimura S. 1976. Multiple host-specific toxins of *Alternaria mali* and their effect on the permeability of host cells. Physiol Plant Pathol. 8(2):141–153.

McMillan VE, Smith RJF. 1974. Agonistic and reproductive behaviour of the fathead minnow (*Pinephales promelas* Rafinesque). Z Tierpsychol. 34(1):25–58.

Morris D. 1967. The naked ape. Toronto: Bantam. 366 p.

Nagei M, Oshima N, Fujii R. 1986. A comparative study of melanin-concentrating hormone (MCH) action on teleost melanophores. Biol Bull. 171(2):360–370.

Young CM, Greenwood PG, Powell CJ. 1986. The ecological role of defensive secretions in the intertidal pulmonate (*Onchidella borealis*). Biol Bull. 171(2):391–404.

Index

Symbols and Abbreviations Used in Biology

TERM/UNIT OF MEASUREMENT	SYMBOL/ ABBREVIATION
ångström	Å
approximately	ca. *or* c. *or* $\approx$
calorie	cal
centimeter	cm
cubic centimeter	cm^3
cubic meter	m^3
cubic millimeter	mm^3
day	d
degree Celsius	°C
degree Fahrenheit	°F
degrees of freedom	df, ν
diameter	diam
east	E
et alia (Latin: "and others")	et al.
et cetera (Latin: "and so forth")	etc.
exempli gratia (Latin: "for example")	e.g.
female	♀
figure, figures	Fig., Figs.
foot-candle	fc *or* ft–c
gram	g
greater than	>
hectare	ha
height	ht
hour	h
id est (Latin: "that is")	i.e.
joule	J
kelvin	K
kilocalorie	Kcal
kilogram	kg
kilometer	km
latitude	lat.
less than	<
liter	L *or* l
logarithm (base 10)	log
logarithm (base *e*)	ln
longitude	long.
male	♂
maximum	max
mean (sample)	$\bar{x}, \bar{X}, \bar{y}, or \bar{Y}$